LECTURES ON HERMITE
AND LAGUERRE EXPANSIONS

Lectures on

HERMITE AND LAGUERRE EXPANSIONS

by

Sundaram Thangavelu

Mathematical Notes 42

PRINCETON UNIVERSITY PRESS

PRINCETON, NEW JERSEY
1993

Copyright © 1993 by Princeton University Press
ALL RIGHTS RESERVED

The Princeton Mathematical Notes are edited by Luis A. Caffarelli,
John N. Mather, and Elias M. Stein

Princeton University Press books are printed on acid-free paper and meet the
guidelines for permanence and durability of the Committee on Production
Guidelines for Book Longevity of the Council on Library Resources

Printed in the United States of America

Library of Congress Cataloging-in-Publication Data

Thangavelu, Sundaram.
    Lectures on Hermite and Laguerre expansions / by Sundaram
Thangavelu.
    p. cm. — (Mathematical notes ; 42)
  Includes bibliographical references.
  ISBN 0-691-00048-4
  1. Hermite polynomials. 2. Laguerre polynomials.
3. Representations of groups. I. Title. II. Series: Mathematical notes
(Princeton University Press) ; 42.
QA404.5.T37   1993
515'.5—dc20   93-16643

Dedicated

To the memory of my father

Veppampalayam Palaniappan Sundaram

Seeker of beauty!
Take a walk through this tiny garden of mine
Curse me not, if you see no flowers
But bear in mind: there are times when
Beauty sleeps on a blade of grass.

CONTENTS

Preface ix

Acknowledgements xv

Chapter 1. Hermite, Special Hermite and Laguerre Functions
§1.1 Hermite and Laguerre Functions 1
§1.2 Heisenberg Group and Weyl Transform 9
§1.3 Special Hermite Functions 14
§1.4 Hermite and Laguerre Functions on the Heisenberg Group 22
§1.5 Asymptotic Properties of Hermite and Laguerre Functions 26

Chapter 2. Special Hermite Expansions
§2.1 Special Hermite Expansions 29
§2.2 Oscillatory Singular Integrals and Riesz Transforms 32
§2.3 Littlewood–Paley–Stein Theory 39
§2.4 Multipliers for Special Hermite Expansions 45
§2.5 Riesz Means for Special Hermite Expansions 51
§2.6 Proof of the Restriction Theorem 59

Chapter 3. Multiple Hermite Expansions
§3.1 Riesz Means and the Critical Index 65
§3.2 An L^2 Estimate for the Riesz Kernel 69
§3.3 Almost Everywhere and Mean Summability 77
§3.4 Hermite Expansions for Radial Functions 81

Chapter 4. Multipliers for Hermite Expansions
§4.1 Littlewood–Paley–Stein Theory for the Hermite Semigroup 85
§4.2 Marcinkiewicz Multiplier Theorem 90
§4.3 Conjugate Poisson Integrals and Riesz Transforms 94
§4.4 Wave Equation for the Hermite Operator 105

Chapter 5. Hermite Expansions on $\mathbb{R}$
§5.1 Cesàro Means and the Critical Index 111
§5.2 An Expression for the Cesàro Kernel and the Basic Estimates 114
§5.3 Estimating the Oscillatory Integrals 119
§5.4 Almost Everywhere and Mean Summability Results 131

Chapter 6. Laguerre Expansions

§6.1 The Laguerre Convolution 137
§6.2 Laguerre Expansions of Convolution Type 143
§6.3 Standard Laguerre Expansions 150
§6.4 Laguerre Expansions of Hermite Type 156

Chapter 7. The Transplantation Theorems

§7.1 The Transplantation Operators 167
§7.2 Further Reduction in the Proof of the Propositions 171
§7.3 Some Preliminary Lemmas 176
§7.4 Estimating $\widetilde{T}_\alpha^{\alpha+i\theta} f$ 181
§7.5 Estimating $\widetilde{T}_\alpha^{\alpha+2+i\theta} f$ 186

Bibliography 193

PREFACE

One of the great pleasures of mathematics is to witness the return of the grand and noble themes from the past. Important insights never die; they may fade from prominence, lose their luster, and be momentarily forgotten amid the swirl of changing fashion; but they remain with us, always, ready to emerge again when the time is ripe, to take their rightful place among the central theories of our day.

There is no better example than the theory of special functions. Forged by the great analysts of the 18th and 19th century in their attempts to solve the differential equations of mathematical physics, these peculiar personalities with their incredibly detailed structure and almost magical properties must have seemed comically old-fashioned and, well yes, special!, in the brave new world of abstract mathematics that swept in during the 20th century. But these were not somebody's pampered pets or idiosynchratic inventions on the fringes of mathematics. These special functions were the irreducibly important personalities, the key characters in so many important dramas that were yet to unfold. They may not have commanded center stage, but they did not go away. They could not be made to disappear by any wave of the wand of generalization because they lay at the heart of the matter for so many important questions. Today we can see that they are back in full force, emerging in so many different ways in the core of harmonic analysis and the representation theory of Lie groups. They arise in almost every attempt to understand examples. General results such as Peter-Weyl theorem, the Plancherel theorem, the theory of interpolation of linear operators, only attain their true significance when understood in the context of specific examples. The examples are not merely illustrations of the general result; they explain the result, the significance of various hypotheses and conclusions, the way it succeeds (and sometimes fails) to illuminate the essential features of the mathematical landscape.

The theory of group representations has been fully developed in three general contexts: 1) locally compact abelian groups, 2) semisimple Lie groups, 3) nilpotent groups. In all the three cases, when we look for the kind of detailed knowledge that reveals more than a superficial understanding, we are forced to return to certain basic examples, notably 1) Euclidean space, 2) rotation and Lorentz groups (more generally, rank one symmetric spaces), 3) the Heisenberg group. Why? In all three cases the basic functions (spherical functions, or more generally entry functions of the irreducible representations) are expressed in terms of classical special functions. For Euclidean space it is Bessel functions. For the sphere it is the Gegenbauer polynomials, for noncompact rank one symmetric spaces it is Legendre and Jacobi functions. For the Heisenberg group it is the Laguerre and Hermite polynomials. The interplay between the properties of Bessel functions and the Euclidean Harmonic analysis (and Gegenbauer polynomials and spherical harmonics) is beautifully described in Stein and Weiss [SW]. A similar exposition for Jacobi functions and rank one symmetric spaces is given in Koornwinder (Jacobi functions and analysis on noncompact semisimple Lie groups, in *Special Functions: Group theoretic aspects and applications*, R. Askey, T. Koornwinder and W. Schempp, Eds., Reidel, Dordrecht, 1984). The story of the Heisenberg group can be found in Folland [Fo] and is further developed in this monograph.

The Hermite functions arise naturally as the eigenfunctions of the harmonic oscillator Hamiltonian, and so play a vital role in quantum physics. They are also the eigenfunctions of the Fourier transform, a point of view that is central in Wiener's development of the Plancherel formula (see N. Wiener, *The Fourier integrals and certain of its applications*, Cambridge (1967)). Multiple (n-dimensional) Hermite functions are defined simply by taking the tensor product of the one-dimensional Hermite functions (just as the n-dimensional Gaussian is constructed from the one-dimensional Gaussians). One might be led to believe therefore that the story of multiple Hermite expansions should follow the same outline as the one-dimensional case; but the reader will discover here a remarkable surprise: the n-dimensional theory is more straightforward, while the one-dimensional theory has rather distinct and deeper features.

The Laguerre polynomials may be regarded as generalizations of the Hermite polynomials in the sense that Laguerre polynomials of order $-1/2$ give the Hermite polynomials of even degree and those of order $1/2$ give the Hermite polynomials of odd degree (see (1.1.52) and (1.1.53) for exact relationship). But there is a deeper connection related to the Weyl transform and the representation theory of the Heisenberg group, which forms a major theme of this work.

The Weyl transform W takes L^2 functions on $\mathbb{C}^n$ to bounded operators on $L^2(\mathbb{R}^n)$ via the identity

$$W(f)\varphi(\xi) = \int_{\mathbb{C}^n} e^{i(x\cdot\xi + \frac{1}{2}x\cdot y)} \varphi(\xi + y) f(x + iy) dx dy \tag{1}$$

for $\varphi \in L^2(\mathbb{R}^n)$, $\xi \in \mathbb{R}^n$. The Heisenberg group is $\mathbb{C}^n \times \mathbb{R}^n$ endowed with the product $(z,t) \cdot (w,s) = (z+w, t+s+\frac{1}{2}\mathrm{Im} z \cdot \overline{w})$. The infinite dimensional irreducible unitary representations of the Heisenberg group are parametrized by a non–zero real λ and are given by

$$\pi_\lambda(z,t)\varphi(\xi) = e^{i\lambda(x\cdot\xi + \frac{1}{2}x\cdot y)} \varphi(\xi + y) \tag{2}$$

acting on the Hilbert space $L^2(R^n)$. Thus

$$W(f)\varphi(\xi) = \int_{\mathbb{C}^n} \pi_1(z,0)\varphi(\xi) f(z) dz. \tag{3}$$

If we choose a basis for $L^2(\mathbb{R}^n)$, then the representation $\pi_\lambda(z,t)$ can be thought of as an infinite matrix, and the entries of this matrix are the entry functions associated with the representation. It is natural to use the Hermite functions Φ_μ on $\mathbb{R}^n$ for the basis. Then the entry functions for π_1 have the form $\Phi_{\mu\nu}(z)e^{it}$ where

$$\Phi_{\mu\nu}(z) = (2\pi)^{-\frac{n}{2}} \int_{\mathbb{R}^n} e^{ix\cdot\xi} \Phi_\mu(\xi + \tfrac{1}{2}y) \Phi_\nu(\xi - \tfrac{1}{2}y) d\xi. \tag{4}$$

These are actually Hermite functions on $\mathbb{C}^n$ viewed as $\mathbb{R}^{2n}$ because

$$(-\Delta + \tfrac{1}{4}|z|^2)\Phi_{\mu\nu} = (|\nu| + |\nu| + n)\Phi_{\mu\nu} \tag{5}$$

but they do not give all Hermite functions on $\mathbb{C}^n$, hence we call them special Hermite functions. But they are also expressible in terms of Laguerre functions. For example,

$$\Phi_{\mu\nu}(z) = (2\pi)^{-\frac{\ell}{n}2)} \prod_{j=1}^{n} L_{\mu_j}(\tfrac{1}{2}|z|^2) e^{-\frac{1}{4}|z|^2}. \tag{6}$$

This work deals with three types of expansions: 1) Hermite functions, 2) special Hermite functions, 3) Laguerre functions (this further subdivides into three subcases). For each type of expansion there are two basic questions: a) is the series convergent or summable in the L^p norm or almost everywhere for any L^p function?, b) is a multiplier transform bounded on L^p if the multiplier satisfies a standard Hörmander condition, and if so how many derivatives are required?

If φ_k is any orthonormal basis for $L^2(d\mu)$, then $\Sigma_{k=1}^n (f,\varphi_k)\varphi_k$ converges to f in $L^2(d\mu)$ as N tends to infinity. However if we let p be different from 2 it is not necessarily true that the above partial sums converge to f in any sense for all f in $L^p(d\mu)$. Because the partial sums of such expansions tend to misbehave, it is convenient to introduce summability factors to obtain positive results. For example, the Riesz means of order α are defined by

$$(7) \qquad S_R^\alpha f = \sum \left(1 - \frac{\lambda_k}{R}\right)_+^\alpha (f,\varphi_k)\varphi_k$$

(λ_k is a natural eigenvalue associated to φ_k which depends on the particular expansion considered). This is a finite sum, and the summability question is whether these means converge to f as R tends to infinity either in the norm or almost everywhere. The convergence of the partial sums corresponds to $\alpha = 0$, and as α increases the summability improves. The smallest α for which L^p convergence holds is called the L^p critical index, and for a number of these expansions the critical index is either known or conjectured. (One can also consider Cesàro means, and there is a general result connecting Riesz and Cesàro summability.)

Multiplier transformations may be defined for general expansions by

$$(8) \qquad T_m f = \sum m(\lambda_k)(f,\varphi_k)\varphi_k.$$

Here $m(\lambda)$ is the multiplier. If m is bounded the T_m is bounded on L^2, but to get T_m to be bounded on L^p, p other than 2 requires more. The standard Hörmander condition is $m^{(j)}(\lambda) = O(\lambda^{-j})$ for all derivatives up to order N. There is an abstract Littlewood–Paley theory due to Stein [S2] which guarantees the L^p boundedness for any p between one and infinity if the multiplier satisfies a condition somewhat stronger than the Hörmander condition with N being infinity (for a large class of expansions). However it is of considerable interest to find the optimal value of N (or something close to the optimal value) for each type of expansion, although this usually requires a great effort.

The techniques used to study these questions for Hermite, special Hermite and Laguerre expansions, run the gamut of modern harmonic analysis, especially the Littlewood–Paley theory, oscillatory singular integrals, the method of stationary phase and van der Corput's lemma. Of course, the generating function identities and asymptotic properties of special functions play an important role. For the Laguerre expansions, the key idea is a recent transplantation theorem of Kanjin [K], which relates the L^p properties of Laguerre expansions of different indices. The point is that for integer index, the Laguerre expansions can be related to special Hermite expansions for which the results can be proved using Heisenberg group methods. The results are then transplanted to general Laguerre expansions.

The results described in this work give a coherent account of what might be called the concrete Littlewood–Paley theory for a class of related expansions. It is a rich and detailed story and the arguments are sometimes fearsomely complex. But in the end there is a satisfying unity of the methods employed and a strong point of view emerges.

Of course, this is an actively evolving field, and one anticipates many important results to come. The reader will find many challenging conjectures and open problems and the invitation to work in this exciting field.

Robert S. Strichartz
Cornell University
May 1992

ACKNOWLEDGEMENTS

The seeds were sown about two and a half years ago by Bob Strichartz. They germinated well but there was absolutely no growth as the initial enthusiasm of the gardener in me faded away soon. Last fall I came to teach at Cornell and the enthusiasm got revived again by Strichartz. With his constant encouragement and many helpful comments I managed to grow some plants and here they are, in my tiny garden, waiting for the visitors.

I am extremely grateful to R. Strichartz for every help he has done during the preparation of this monograph. I am also thankful to Eli Stein for his encouragement and suggestions. This monograph is based on my lectures given as a graduate course during the spring semester of 1992 at Cornell. I wish to thank the University for the financial support and my home institute T.I.F.R. for giving me leave. I also wish to thank Ron Kerman for patiently listening to my lectures and making many interesting enquiries.

Though my mumblings were accepted as lectures, I am doubtful if anybody would have accepted my handwritten notes as a respectable monograph. As an old poem says, even God has to be well dressed if he desires to be respected. In this respect I was fortunate to have the golden touch of Mrs. Arletta Havlik transforming the entire manuscript into a beautiful book. I take great pleasure in acknowledging her help and expressing my heartfelt thanks.

Finally, it goes without saying that I am thankful to my wife and daughter who were forced to remain within four walls and watch the melting snow and the nesting birds for more than three months. As the grass emerged triumphant from beneath the ice and the birds started singing their songs, my garden too was ready and once again they got me back all for themselves.

LECTURES ON HERMITE
AND LAGUERRE EXPANSIONS

CHAPTER 1
HERMITE, SPECIAL HERMITE AND LAGUERRE FUNCTIONS

In this chapter we define Hermite, Laguerre and special Hermite functions and prove some of their properties which are needed in studying the expansions in terms of them. In order to derive the special Hermite functions we recall briefly some results from the representation theory of the Heisenberg group. We also define and prove some properties of Weyl transforms which we need in the study of special Hermite expansions. Some important asymptotic properties and norm estimates are collected in Section 1.5.

1.1 Hermite and Laguerre Functions

Hermite functions arise in many contexts and as such there are several ways of defining them. Here we take the most classical definition. Hermite polynomials $H_k(x)$ are defined on the real line by the formula

(1.1.1) $$H_k(x) = (-1)^k \frac{d^k}{dx^k}(e^{-x^2})e^{x^2}, \quad k = 0, 1, 2, \ldots.$$

We then define the Hermite functions $\widetilde{h}_k$ by setting

(1.1.2) $$\widetilde{h}_k(x) = H_k(x)e^{-\frac{1}{2}x^2}, \quad k = 0, 1, 2, \ldots.$$

Many properties of the Hermite functions follow directly from the above definition. We record here some properties which are needed in the sequel.

First of all we have the following generating function identity for the Hermite polynomials. If $|w| < 1$ then we have

(1.1.3) $$\sum_{k=0}^{\infty} \frac{H_k(x)}{k!} w^k = e^{2xw - w^2}.$$

This can be easily proved by Taylor expanding the function on the right hand side about $w = 0$ and making use of the definition (1.1.1) in calculating the derivatives. From the generating function (1.1.3) we obtain the following relations.

(1.1.4) $$H_k'(x) = 2k H_{k-1}(x), \quad H_k(x) = 2x H_{k-1}(x) - H_{k-1}'(x).$$

For the Hermite functions $\widetilde{h}_k$ these relations take the form

(1.1.5) $$\left(-\frac{d}{dx} + x\right)\widetilde{h}_k(x) = \widetilde{h}_{k+1}(x),$$

and

(1.1.6) $$\left(\frac{d}{dx} + x\right)\widetilde{h}_k(x) = 2k\widetilde{h}_{k-1}(x).$$

The operators $A = -\frac{d}{dx} + x$ and $A^* = \frac{d}{dx} + x$ are called the creation and annihilation operators in quantum mechanics. The formulas (1.1.5) and (1.1.6) give the recursion relation

(1.1.7) $$2x\widetilde{h}_k(x) = \widetilde{h}_{k+1}(x) + 2k\widetilde{h}_{k-1}(x).$$

These formulas will play an important role in the study of Hermite expansions.

Now, an easy calculation shows that $H = -\frac{d^2}{dx^2} + x^2$, the Hermite operator, can be written in the form

(1.1.8) $$H = \tfrac{1}{2}(AA^* + A^*A).$$

The Hermite functions $\widetilde{h}_k$ are then eigenfunctions of this operator. In fact, the relations (1.1.5), (1.1.6) and (1.1.8) immediately give us

(1.1.9) $$H(\widetilde{h}_k) = (2k+1)\widetilde{h}_k.$$

This also proves that the functions $\widetilde{h}_k$ form an orthogonal family in $L^2(\mathbb{R}, dx)$. This can be proved by integrating the relation

(1.1.10) $$2(k-j)\widetilde{h}_k(x)\widetilde{h}_j(x) = \widetilde{h}_k(x)\widetilde{h}_j''(x) - \widetilde{h}_j(x)\widetilde{h}_k''(x)$$

which follows from (1.1.9). We want to suitably normalize them so that they will form an orthonormal family. To that end we prove another important generating function identity, also known as Mehler's formula, for the Hermite functions.

Lemma 1.1.1. *For $|w| < 1$ we have*

(1.1.11) $$\sum_{k=0}^{\infty} \frac{\widetilde{h}_k(x)\widetilde{h}_k(y)}{2^k k!} w^k = (1-w^2)^{-\frac{1}{2}} e^{-\frac{1}{2}\frac{1+w^2}{1-w^2}(x^2+y^2) + \frac{2w}{1-w^2}xy}.$$

Proof: We start with the well known formula

(1.1.12) $$e^{-x^2} = \frac{1}{\sqrt{\pi}} \int_{-\infty}^{\infty} e^{-u^2 + 2ixu} du.$$

From this we obtain the formula

(1.1.13) $$\widetilde{h}_k(x) = \frac{1}{\sqrt{\pi}} (-2i)^k e^{\frac{1}{2}x^2} \int_{-\infty}^{\infty} u^k e^{-u^2 + 2ixu} du,$$

and hence we have

(1.1.14) $$\sum_{k=0}^{\infty} \frac{\widetilde{h}_k(x)\widetilde{h}_k(y)}{2^k k!} w^k$$

$$= \frac{1}{\pi} e^{\frac{1}{2}(x^2+y^2)} \sum_{k=0}^{\infty} \int_{-\infty}^{\infty}\int_{-\infty}^{\infty} \frac{(-2uvw)^k}{k!} e^{-u^2-v^2+2ixu+2iyv} \, du dv$$

$$= \frac{1}{\pi} e^{\frac{1}{2}(x^2+y^2)} \int_{-\infty}^{\infty}\int_{-\infty}^{\infty} e^{-u^2-v^2+2ixu+2iyv-2uvw} \, du dv$$

$$= \frac{1}{\sqrt{\pi}} e^{\frac{1}{2}(x^2-y^2)} \int_{-\infty}^{\infty} e^{-(1-w^2)u^2 + 2i(x-yw)u} du$$

$$= (1-w^2)^{-\frac{1}{2}} e^{\frac{1}{2}(x^2-y^2)} e^{-\frac{(x-yw)^2}{1-w^2}}.$$

This completes the proof of the lemma. ∎

Lemma 1.1.2.

(1.1.15) $$\int_{-\infty}^{\infty} (\widetilde{h}_k(x))^2 dx = 2^k k! \sqrt{\pi}.$$

Proof: Taking $x = y$ in formula (1.1.11) we get

(1.1.16) $$\sum_{k=0}^{\infty} \frac{(\widetilde{h}_k(x))^2}{2^k k!} w^k = (1-w^2)^{-\frac{1}{2}} e^{-\frac{1-w}{1+w} x^2}.$$

Integrating both sides we see that

(1.1.17) $$\sum_{k=0}^{\infty} \left(\int_{-\infty}^{\infty} (\widetilde{h}_k(x))^2 dx \right) \frac{w^k}{2^k k!} = (1-w^2)^{-\frac{1}{2}} \int_{-\infty}^{\infty} e^{-\frac{1-w}{1+w} x^2} dx$$

$$= \sqrt{\pi}(1-w^2)^{-\frac{1}{2}} \left(\frac{1+w}{1-w} \right)^{\frac{1}{2}}$$

$$= \sqrt{\pi} \sum_{k=0}^{\infty} w^k.$$

Equating the coefficients on both sides we obtain the Lemma. ∎

Therefore, we define the normalized Hermite functions $h_k(x)$ by

(1.1.18) $$h_k(x) = (2^k k! \sqrt{\pi})^{-\frac{1}{2}} \tilde{h}_k(x).$$

Then they form an orthonormal family in $L^2(\mathbb{R}, dx)$. It can be shown that this family is complete. Later when we estimate certain kernels defined by Hermite functions we need to know the exact values of $h_k(0)$. These values can be calculated from Mehler's formula. The generating function (1.1.3) shows that $H_k(-x) = (-1)^k H_k(x)$ so that h_{2k} is even and h_{2k+1} is odd. Therefore, $h_{2k+1}(0) = 0$ and the values $h_{2k}(0)$ can be calculated from

(1.1.19) $$\sum_{k=0}^{\infty} (h_{2k}(0))^2 w^{2k} = \pi^{-\frac{1}{2}} (1 - w^2)^{-\frac{1}{2}}.$$

If we use the expansion

(1.1.20) $$(1 - w^2)^{-\frac{1}{2}} = \sum_{k=0}^{\infty} \frac{\Gamma(k + \frac{1}{2})}{\Gamma(k + 1) \sqrt{\pi}} w^{2k}$$

it follows that

(1.1.21) $$(h_{2k}(0))^2 = \frac{1}{\pi} \frac{\Gamma(k + \frac{1}{2})}{\Gamma(k + 1)}.$$

We also need several asymptotic properties and estimates for the Hermite functions. These will be stated in Section 1.5.

We have seen that the Hermite functions are eigenfunctions of the operator H. It is interesting to note that they are also eigenfunctions of the Fourier transform. This is not surprising since H is invariant under the Fourier transform. We take the definition of the Fourier transforms to be

(1.1.22) $$\hat{f}(\xi) = \frac{1}{\sqrt{2\pi}} \int_{-\infty}^{\infty} e^{-ix \cdot \xi} f(x) dx.$$

From this definition it is easy to see that

(1.1.23) $$(xf)\widehat{\,}(\xi) = i \frac{\partial}{\partial \xi} \hat{f}(\xi), \quad \left(-\frac{d}{dx} f \right)\widehat{\,}(\xi) = -i\xi \hat{f}(\xi).$$

We now prove the following result.

Lemma 1.1.3. *The Hermite functions are eigenfunctions of the Fourier transform:*
$$\widehat{h}_k(\xi) = (-i)^k h_k(\xi).$$

Proof: The formula (1.1.5) gives us

(1.1.24) $$\left(-\frac{d}{dx} + x\right) h_k(x) = (2k+2)^{\frac{1}{2}} h_{k+1}(x).$$

If we take Fourier transform on both sides and use (1.1.23) we get

(1.1.25) $$-i\left(-\frac{\partial}{\partial \xi} + \xi\right) \widehat{h}_k(\xi) = (2k+2)^{\frac{1}{2}} \widehat{h}_{k+1}(\xi).$$

If we assume that the lemma is true for h_k then it follows that

(1.1.26) $$\begin{aligned}(2k+2)^{\frac{1}{2}} \widehat{h}_{k+1}(\xi) &= (-i)^{k+1}\left(-\frac{\partial}{\partial \xi} + \xi\right) h_k(\xi) \\ &= (-i)^{k+1}(2k+2)^{\frac{1}{2}} h_{k+1}(\xi).\end{aligned}$$

Therefore, it is enough to show that $\widehat{h}_0(\xi) = h_0(\xi)$. But $h_0(x) = \pi^{-\frac{1}{4}} e^{-\frac{1}{2}x^2}$ and hence $\widehat{h}_0 = h_0$ is immediate. ∎

We now define Hermite functions on $\mathbb{R}^n$. Let μ be a multiindex and $x \in \mathbb{R}^n$. $\Phi_\mu(x)$ is defined by taking the product of the one dimensional Hermite functions $h_{\mu_j}(x_j)$:

(1.1.27) $$\Phi_\mu(x) = \prod_{j=1}^n h_{\mu_j}(x_j).$$

Then they form a complete orthonormal system for $L^2(\mathbb{R}^n, dx)$. Many properties of Φ_μ follow from the corresponding properties of the one dimensional functions. For example, if $H = -\Delta + |x|^2$ is the Hermite operator on $\mathbb{R}^n$, then

(1.1.28) $$H\Phi_\mu = (2|\mu| + n)\Phi_\mu$$

where $|\mu| = \mu_1 + \mu_2 + \cdots + \mu_n$. The operator H can be written in the form

(1.1.29) $$H = \tfrac{1}{2} \sum_{j=1}^n (A_j A_j^* + A_j^* A_j)$$

where $A_j = -\frac{\partial}{\partial x_j} + x_j$ and $A_j^* = \frac{\partial}{\partial x_j} + x_j$. In view of the formula (1.1.5) and (1.1.6) it follows that

$$(1.1.30) \qquad A_j \Phi_\mu = (2\mu_j + 2)^{\frac{1}{2}} \Phi_{\mu+e_j}, \quad A_j^* \Phi_\mu = (2\mu_j)^{\frac{1}{2}} \Phi_{\mu-e_j}$$

where e_j are the coordinate vectors.

Given a function f on $\mathbb{R}^n$ we define the Fourier–Hermite coefficients $\widehat{f}(\mu)$ by

$$(1.1.31) \qquad \widehat{f}(\mu) = \int_{\mathbb{R}^n} f(x) \Phi_\mu(x) dx.$$

The Hermite expansion of f is then given by

$$(1.1.32) \qquad f(x) = \sum_\mu \widehat{f}(\mu) \Phi_\mu(x).$$

This expansion is formal and need not converge. Our main concern in Chapters 3 and 5 is to study the convergence properties of the above expansion. Defining P_k to be the projection onto the kth eigenspace

$$(1.1.33) \qquad P_k f = \sum_{|\mu|=k} \widehat{f}(\mu) \Phi_\mu$$

we can write (1.1.32) in the form

$$(1.1.34) \qquad f = \sum_{k=0}^\infty P_k f.$$

The projections are integral operators with kernels

$$(1.1.35) \qquad \Phi_k(x,y) = \sum_{|\mu|=k} \Phi_\mu(x) \Phi_\mu(y).$$

From Lemma 1.1.1 and the definition of Φ_μ it follows that

$$(1.1.36) \qquad \sum_{k=0}^\infty r^k \Phi_k(x,y) = \pi^{-\frac{n}{2}} (1-r^2)^{-\frac{n}{2}} e^{-\frac{1}{2}\frac{1+r^2}{1-r^2}(|x|^2+|y|^2) + \frac{2rx \cdot y}{1-r^2}}.$$

This is the n-dimensional version of the Mehler's formula and is very useful in the study of Hermite expansions.

Next we consider Laguerre polynomials. For $\alpha > -1$, Laguerre polynomials of type α are defined by the formula

$$(1.1.37) \qquad e^{-x} x^\alpha L_k^\alpha(x) = \frac{1}{k!} \frac{d^k}{dx^k} (e^{-x} x^{k+\alpha}).$$

Here $x > 0$ and $k = 0, 1, 2, \ldots$. Each L_k^α is a polynomial of degree k; it is explicitly given by

$$(1.1.38) \qquad L_k^\alpha(x) = \sum_{j=0}^{k} \frac{\Gamma(k+\alpha+1)}{\Gamma(k-j+1)\Gamma(j+\alpha+1)} \frac{(-x)^j}{j!}.$$

When $n = 0$ we have

$$(1.1.39) \qquad L_k^\alpha(0) = \frac{\Gamma(k+\alpha+1)}{\Gamma(k+1)\Gamma(\alpha+1)}.$$

Lemma 1.1.4. *The Laguerre polynomials satisfy the orthogonality properties*

$$\int_0^\infty L_k^\alpha(x) L_j^\alpha(x) e^{-x} x^\alpha \, dx = \frac{\Gamma(k+\alpha+1)}{\Gamma(k+1)} \delta_{jk}.$$

Proof: Let f be any polynomial and consider

$$(1.1.40) \qquad \int_0^\infty f(x) L_k^\alpha(x) e^{-x} x^\alpha \, dx = \frac{1}{k!} \int_0^\infty f(x) \frac{d^k}{dx^k} (e^{-x} x^{k+\alpha} \, dx).$$

Integrating by parts we see that

$$(1.1.41) \qquad \int_0^\infty f(x) L_k^\alpha(x) e^{-x} x^\alpha \, dx = \frac{(-1)^k}{k!} \int_0^\infty f^{(k)}(x) e^{-x} x^{k+\alpha} \, dx.$$

If f is a polynomial of degree $j < k$ then it follows that

$$\int_0^\infty f(x) L_k^\alpha(x) e^{-x} x^\alpha \, dx = 0.$$

In particular this proves that when $k \neq j$

$$(1.1.42) \qquad \int_0^\infty L_k^\alpha(x) L_j^\alpha(x) e^{-x} x^\alpha \, dx = 0.$$

And when $k = j$, taking $f(x) = L_k^\alpha(x)$ we observe that $f^{(k)}(x) = (-1)^k$ so that

$$(1.1.43) \qquad \int_0^\infty L_k^\alpha(x) L_k^\alpha(x) e^{-x} x^\alpha \, dx = \frac{1}{k!} \int_0^\infty x^{k+\alpha} e^{-x} \, dx$$

which proves the Lemma. ∎

Therefore, if we define $\mathcal{L}_k^\alpha(x)$ by

$$(1.1.44) \qquad \mathcal{L}_k^\alpha(x) = \left(\frac{\Gamma(k+1)}{\Gamma(k+\alpha+1)} \right)^{\frac{1}{2}} e^{-x} x^{\frac{\alpha}{2}} L_k^\alpha(x)$$

then it follows that they form an orthonormal family in $L^2(\mathbb{R}_+, dx)$ where $\mathbb{R}_+ = (0, \infty)$. To end this section we collect some formulas satisfied by the Laguerre functions. The following three generating function identities are useful in the study of Laguerre expansions. In what follows we assume $|w| < 1$.

$$(1.1.45) \qquad \sum_{k=0}^\infty L_k^\alpha(x) w^k = (1-w)^{-\alpha-1} e^{-\frac{w}{1-w} x},$$

$$(1.1.46) \qquad \sum_{k=0}^\infty \frac{L^\alpha(x)}{\Gamma(k+\alpha+1)} w^k = e^w (xw)^{-\frac{\alpha}{2}} J_\alpha(2(xw)^{\frac{1}{2}}),$$

$$(1.1.47) \qquad \sum_{k=0}^\infty \frac{\Gamma(k+1)}{\Gamma(k+\alpha+1)} L_k^\alpha(x) L_k^\alpha(y) w^k$$
$$= (1-w)^{-1} (-xyw)^{-\frac{\alpha}{2}} e^{-\frac{w}{1-w}(x+y)} J_\alpha \left(\frac{2(-xyw)^{\frac{1}{2}}}{1-w} \right).$$

In the above formulas J_α stands for the Bessel function of order α.

The Laguerre polynomials $L_k^\alpha(x)$ satisfy the differential equation

$$(1.1.48) \qquad xy''(x) + (\alpha + 1 - x) y'(x) + k y(x) = 0$$

and one has the relation

$$(1.1.49) \qquad \frac{d}{dx} L_k^\alpha(x) = -L_{k-1}^{\alpha+1}(x).$$

They also satisfy the recursion relation

$$(1.1.50) \quad k L_k^\alpha(x) = (-x + 2k + \alpha - 1) L_{k-1}^\alpha(x) - (k + \alpha - 1) L_{k-2}^\alpha(x)$$

and Laguerre polynomials of different order are connected by the formula

$$(1.1.51) \qquad \sum_{j=0}^{k} A_{k-j}^{\alpha} L_j^{\beta}(x) = L_k^{\alpha+\beta+1}(x).$$

Here $A_k^{\alpha} = \dfrac{\Gamma(k+\alpha+1)}{\Gamma(k+1)\Gamma(\alpha+1)}$ are the binomial coefficients. We conclude this section with certain formulas connecting Hermite and Laguerre polynomials. The Hermite polynomials satisfy

$$(1.1.52) \qquad H_{2k}(x) = (-1)^k 2^{2k} k! L_k^{-\frac{1}{2}}(x^2),$$

and

$$(1.1.53) \qquad H_{2k+1}(x) = (-1)^k 2^{2k} k! L_k^{\frac{1}{2}}(x^2) x.$$

If $\alpha > -\frac{1}{2}$, then the Laguerre polynomials can be expressed in terms of H_{2k} as follows:

$$(1.1.54) \quad L_k^{\alpha}(x) = \frac{(-1)^k \pi^{-\frac{1}{2}}}{\Gamma(\alpha+\frac{1}{2})} \frac{\Gamma(k+\alpha+1)}{(2k)!} \int_{-1}^{1} (1-t^2)^{\alpha-\frac{1}{2}} H_{2k}(\sqrt{x}t) dt.$$

This formula will be used in the study of special Hermite expansions.

1.2 Heisenberg Group and Weyl Transform

Analysis on the Heisenberg group and expansions in terms of Hermite and Laguerre functions are interrelated. On the one hand it became clear from the work of Geller [Ge] that harmonic analysis on the Heisenberg group heavily depends on many properties of Hermite and Laguerre functions. On the other hand, analysis on the Heisenberg group also plays an important role in the study of Hermite and Laguerre expansions. In fact, the first summability theorem for multiple Hermite expansions was deduced from the corresponding result on the Heisenberg group by Hulanicki and Jenkins [HJ]. A multiplier theorem for Laguerre expansions was likewise proved using a multiplier theorem on the Heisenberg group, by Długosz [Dl]. The first multiplier theorem for Hermite expansions, in the work of Mauceri [M] followed from considerations of the Weyl transform which is related to the Schrödinger representation π_1 on the Heisenberg group.

More interestingly, the special Hermite functions, which occupy a central place in the study of Hermite and Laguerre expansions are defined in terms of the Schrödinger representation on the Heisenberg group. The Weyl transform and the twisted convolution which are close relatives of the Fourier transform and the convolution on the Heisenberg group, play a significant role in our study. Our aim in this section is therefore to review briefly the representation theory on the Heisenberg group and to show how various objects of interest arise from considerations on the Heisenberg group. For a readable account of many properties of the Heisenberg group we refer to Folland [Fo] and Taylor [Ta].

To begin with we define the Heisenberg group H^n to be a nilpotent Lie group whose underlying manifold is $\mathbb{C}^n \times \mathbb{R}$. The group law is defined by

$$(1.2.1) \qquad (z,t) \cdot (w,s) = \left(z + w, t + s + \tfrac{1}{2} \operatorname{Im} z \cdot \overline{w}\right).$$

Under this multiplication H^n becomes a nilpotent unimodular Lie group, the Haar measure being the Lebesgue measure $dz\, dt$ on $\mathbb{C}^n \times \mathbb{R}$. The corresponding Lie algebra h_n is generated by the $(2n+1)$ left invariant vector fields

$$(1.2.2) \qquad X_j = \left(\frac{\partial}{\partial x_j} - \frac{1}{2} y_j \frac{\partial}{\partial t}\right), \quad j = 1, 2, \ldots, n$$

$$(1.2.3) \qquad Y_j = \left(\frac{\partial}{\partial y_j} + \frac{1}{2} x_j \frac{\partial}{\partial t}\right), \quad j = 1, 2, \ldots, n$$

and $T = i \frac{\partial}{\partial t}$. The operator

$$(1.2.4) \qquad \mathcal{L} = -\sum_{j=1}^{n}(X_j^2 + Y_j^2)$$

is called the sublaplacian (Heisenberg Laplacian or Kohn Laplacian) which is subelliptic but not elliptic.

The representation theory on the Heisenberg group is fairly simple and well understood. There are two types of representations. The infinite dimensional representations are parametrized by nonzero reals. The one dimensional representations are not important to us and so we will not discuss them. For each $\lambda \neq 0$, there is an infinite dimensional representation $\pi_\lambda(z,t)$ which in the Schrödinger realization acts on $L^2(\mathbb{R}^n)$ in the following way. For each $\varphi \in L^2(\mathbb{R}^n)$

$$(1.2.5) \qquad \pi_\lambda(z,t)\varphi(\xi) = e^{i\lambda t} e^{i\lambda(x \cdot \xi + \frac{1}{2} x \cdot y)} \varphi(\xi + y).$$

It is easily seen that

(1.2.6) $$\pi_\lambda(z,t)\pi_\lambda(w,s) = \pi_\lambda((z,t)(w,s))$$

and $\pi_\lambda(z,t)$ is a unitary operator on $L^2(\mathbb{R}^n)$. In other words, $\pi_\lambda(z,t)$ is a unitary representation of H^n on $L^2(\mathbb{R}^n)$. In the next section we show that $\pi_\lambda(z,t)$ are all irreducible. Up to unitary equivalence these give all the infinite dimensional irreducible unitary representations. That is the theorem of Stone and von Neumann.

We now proceed to define the Weyl transform and the twisted convolution. The Schrödinger representation π_1 of H^n is not faithful. The kernel of π_1 is the subgroup

(1.2.7) $$\Gamma = \{(0,k) : k \in \mathbb{Z}\}.$$

For some purposes it may be better to throw away this kernel which leads to the definition of the reduced Heisenberg group H^n_{red}. This is defined to be the quotient H^n/Γ. The elements of H^n/Γ are still written as (z,t) with the understanding that $0 \leq t < 2\pi$. Then π_1 is a representation of H^n_{red} which is now faithful. Since the central variable t acts in a simple minded way it is convenient to disregard it. We define $W(z) = \pi_1(z,0)$ so that $\pi_1(z,t) = e^{it}W(z)$. This $W(z)$ then defines a projective representation of $\mathbb{C}^n$ on the Hilbert space $L^2(\mathbb{R}^n)$.

The unitary representation π_1 of H^n/Γ also defines a representation of the algebra $L^1(H^n/\Gamma)$ in the following way. For f in $L^1(H^n/\Gamma)$, $\pi_1(f)$ is defined to be the operator

(1.2.8) $$\pi_1(f) = \int_{H^n/\Gamma} f(z,t)\pi_1(z,t)dzdt.$$

If $f * g$ is the convolution of f and g on H^n/Γ defined by

(1.2.9) $$f * g(z,t) = \int_{H^n/\Gamma} f((z,t) \cdot (w,s)^{-1})g(w,s)dwds$$

then we have the relation

(1.2.10) $$\pi_1(f * g) = \pi_1(f)\pi_1(g).$$

Since $\pi_1(z,t) = e^{it}W(z)$ we can write $\pi_1(f)$ as

(1.2.11) $$\pi_1(f) = \sqrt{2\pi}\int_{\mathbb{C}^n} f_1(z)W(z)dz$$

where $f_1(z)$ is the Fourier coefficient

$$(1.2.12) \qquad f_1(z) = \frac{1}{\sqrt{2\pi}} \int_0^{2\pi} f(z,t) e^{it} dt.$$

Given a function f in $L^1(\mathbb{C}^n)$, the function $f^\#(z,t) = (2\pi)^{-1} e^{-it} f(z)$ belongs to $L^1(H^n/\Gamma)$ and one has the relation

$$(1.2.13) \qquad \pi_1(f^\#) = \int_{\mathbb{C}^n} f(z) W(z) dz.$$

This transform which takes f into the operator $W(f) = \pi_1(f^\#)$ is called the Weyl transform.

The group convolution on H^n/Γ can be transferred to $\mathbb{C}^n$ as a nonstandard convolution. Given two functions f and g in $L^1(\mathbb{C}^n)$ we calculate $f^\# * g^\#$:

$$(1.2.14) \quad f^\# * g^\#(z,t)$$
$$= \int_{H^n/\Gamma} f^\#(z-w, t-s-\tfrac{1}{2}\operatorname{Im} z\cdot \overline{w}) g^\#(w,s) dw ds$$
$$= (2\pi)^{-1} \left(\int_{\mathbb{C}^n} f(z-w) g(w) e^{\frac{1}{2}\operatorname{Im} z \cdot \overline{w}} dw \right) e^{-it}.$$

Therefore, if we define $f \times g$ by

$$(1.2.15) \qquad f \times g(z) = \int_{\mathbb{C}^n} f(z-w) g(w) e^{\frac{1}{2}\operatorname{Im} z \cdot \overline{w}} dw$$

then we have the relation $f^\# * g^\# = (f \times g)^\#$. This convolution (1.2.15) is called the twisted convolution and $L^1(\mathbb{C}^n)$ becomes an algebra under this. Moreover, as $W(f) = \pi_1(f^\#)$ and π_1 is a representation of the algebra $L^1(H^n/\Gamma)$ we get the relation

$$(1.2.16) \qquad W(f \times g) = W(f) W(g).$$

This means that W defines a representation of the twisted convolution algebra $L^1(\mathbb{C}^n)$.

We now prove a Plancherel theorem and an inversion formula for the Weyl transform. From the definition it follows that $W(f)$ is an integral operator with kernel $K_f(x,y)$ defined by

$$(1.2.17) \qquad K_f(x,y) = \int_{\mathbb{R}^n} f(\xi, y-x) e^{i(x+y)\cdot \xi} d\xi$$

where we have written $f(x,y)$ in place of $f(x+iy)$. Using this explicit form of the kernel we can easily prove the following theorem.

Theorem 1.2.1. *The Weyl transform maps $L^1(\mathbb{C}^n)$ into the space of compact operators on $L^2(\mathbb{R}^n)$. When f is in $L^2(\mathbb{C}^n)$, $W(f)$ is a Hilbert–Schmidt operator and we have the Plancherel formula*

(1.2.18) $$\|f\|_2 = (2\pi)^{-n/2} \|W(f)\|_{HS}.$$

We also have the inversion formula

(1.2.19) $$f(z) = (2\pi)^n \operatorname{tr}(W(z)^* W(f))$$

where $W(z)^*$ is the adjoint of $W(z)$.

Proof: Assume that $f \in L^1(\mathbb{C}^n) \cap L^2(\mathbb{C}^n)$. The Hilbert–Schmidt norm of $W(f)$ is given by

(1.2.20) $$\|W(f)\|_{HS}^2 = \int_{\mathbb{R}^{2n}} |K_f(x,y)|^2 dx dy.$$

Since $K_f(x,y) = \mathcal{F}_1 f(x+y, y-x)$ where $\mathcal{F}_1 f$ stands for the Fourier transform of f in the first variable, a change of variables and Euclidean Plancherel theorem proves that

(1.2.21) $$\|W(f)\|_{HS}^2 = (2\pi)^{-n} \|f\|_2^2.$$

As $f \in L^1 \cap L^2$, we also have

(1.2.22) $$\|W(f)\| \leq \|f\|_1,$$

and $W(f)$ is a compact operator. A density argument proves that $W(f)$ is compact whenever f is in $L^1(\mathbb{C}^n)$. The inversion formula is proved in a similar way by calculating the trace of $W(z)^* W(f)$ and applying the inversion theorem for the Fourier transform. ∎

We conclude this section with some remarks on the twisted convolution. Like the ordinary convolution, twisted convolution also extends from $L^1(\mathbb{C}^n)$ to other $L^p(\mathbb{C}^n)$ and satisfies the Young's inequality

(1.2.23) $$\|f \times g\|_r \leq \|f\|_p \|f\|_q, \quad \frac{1}{r} = \frac{1}{p} + \frac{1}{q} - 1.$$

Though the twisted convolution is not commutative it has better behavior with respect to L^p estimates. Just to give an example we have the following theorem which is not true for ordinary convolution.

Theorem 1.2.2. For f and g in $L^2(\mathbb{C}^n)$, $f \times g$ is also in $L^2(\mathbb{C}^n)$ and $\|f \times g\|_2 \leq (2\pi)^{n/2} \|f\|_2 \|g\|_2$.

Proof: The proof follows from the Plancherel formula (1.2.18) and the relation $W(f \times g) = W(f)W(G)$. In fact, the kernel of $W(f \times g)$ is equal to
$$K_{f \times g}(x, z) = \int K_f(x, y) K_g(y, z) dy$$
and this gives
$$\|W(f \times g)\|_{HS}^2 = \int |K_{f \times g}(x, z)|^2 dx dz$$
$$\leq \left(\int |K_f(x, y)|^2 dx dy \right) \left(\int |K_g(y, z)|^2 dy dz \right)$$
$$= \|W(f)\|_{HS}^2 \|W(g)\|_{HS}^2.$$
Therefore,
$$\|f \times g\|_2^2 = (2\pi)^{-n} \|W(f \times g)\|_{HS}^2$$
$$\leq (2\pi)^{-n} \|W(f)\|_{HS}^2 \|W(g)\|_{HS}^2 \leq (2\pi)^n \|f\|_2^2 \|g\|_2^2 . \quad \blacksquare$$

This theorem will be used in the study of oscillatory singular integrals in Chapter 2.

1.3 Special Hermite Functions

In this section we define and prove some important properties of special Hermite functions. As we have remarked, these functions are defined using the Schrödinger representation π_1. For functions f and g on $\mathbb{R}^n$ we define their Fourier–Wigner transform to be the function

$$(1.3.1) \qquad V(f, g)(z) = (2\pi)^{-\frac{n}{2}} \int_{\mathbb{R}^n} e^{ix \cdot \xi} f\left(\xi + \frac{1}{2}y\right) \overline{g}\left(\xi - \frac{1}{2}y\right) d\xi$$

where $z = (x + iy) \in \mathbb{C}^n$. For this transform we easily prove the following result.

Proposition 1.3.1. For f, g, φ, ψ in $L^2(\mathbb{R}^n)$ one has

$$\int_{\mathbb{C}^n} V(f, g)(z) \overline{V(\varphi, \psi)}(z) dz = (f, \varphi)(\psi, g)$$

where the bracket on the right hand side stands for the inner product in $L^2(\mathbb{R}^n)$.

Proof: Observe that $V(f,g)$ is sesquilinear. Therefore, it is enough to prove that

$$(1.3.2) \qquad \int_{\mathbb{C}^n} |V(f,g)|^2 dz = \|f\|_2^2 \|g\|_2^2.$$

In view of the Plancherel theorem for the Fourier transform, we have

$$(1.3.3) \qquad \int |V(f,g)|^2 dz = \int \left|f\left(\xi + \frac{1}{2}y\right)\right|^2 \left|g\left(\xi - \frac{1}{2}y\right)\right|^2 d\xi dy.$$

A change of variables now proves (1.3.2). ∎

Now we can prove that the representations introduced in the previous section are actually irreducible.

Theorem 1.3.1. *The representations $\pi_\lambda(z,t)$ are irreducible for any $\lambda \neq 0$.*

Proof: Without loss of generality we can assume that $\lambda = 1$. If $M \subset L^2(\mathbb{R}^n)$ is invariant under $\pi_1(z,t)$ for all $(z,t) \in H^n$ and if g is orthogonal to M then

$$(1.3.4) \qquad \int \pi_1(z,t) f(\xi) \overline{g}(\xi) d\xi = 0$$

which is to say that $V(f,g) = 0$. But by the above proposition

$$(1.3.5) \qquad 0 = \|V(f,g)\|_2 = \|f\|_2 \|g\|_2$$

which shows that $g = 0$. Hence $M = L^2(\mathbb{R}^n)$. ∎

We now define the special Hermite functions as the Fourier–Wigner transform of the Hermite functions on $\mathbb{R}^n$. For each pair of multiindices μ and ν we define $\Phi_{\mu\nu}(z) = V(\Phi_\mu, \Phi_\nu)(z)$. This can also be put in the form

$$(1.3.6) \qquad \Phi_{\mu\nu}(z) = (2\pi)^{-\frac{n}{2}} (W(z)\Phi_\mu, \Phi_\nu).$$

We then have the following theorem.

Theorem 1.3.2. *The special Hermite functions $\Phi_{\mu\nu}$ form a complete orthonormal system in $L^2(\mathbb{C}^n)$.*

Proof: The orthonormality follows from Proposition 1.3.1. To prove the completeness assume that $(f, \Phi_{\mu\nu}) = 0$ for all ν and μ. To show that $f = 0$ we calculate $(W(\overline{f})\Phi_\mu, \Phi_\nu)$. From the definition it follows that

$$(1.3.7) \qquad (W(\overline{f})\Phi_\mu, \Phi_\nu) = \int \overline{f}(z)(W(z)\Phi_\mu, \Phi_\nu) dz$$

and hence $(W(\overline{f})\Phi_\mu, \Phi_\nu) = 0$. As the Hermite functions Φ_ν form an orthonormal basis for $L^2(\mathbb{R}^n)$, the above means that $W(\overline{f}) = 0$. But then by the Plancherel theorem for the Weyl transform we get $f = 0$. ∎

We now show that our special Hermite functions are eigenfunctions of a second order elliptic operator L on $\mathbb{C}^n$. To define this operator we introduce the following $2n$ vector fields on $\mathbb{C}^n$.

$$(1.3.8) \qquad Z_j = \left(\frac{\partial}{\partial z_j} + \frac{1}{2}\overline{z}_j\right), \quad j = 1, 2, \ldots, n,$$

$$(1.3.9) \qquad \overline{Z}_j = \left(\frac{\partial}{\partial \overline{z}_j} - \frac{1}{2}z_j\right), \quad j = 1, 2, \ldots, n.$$

These $2n$ vector fields together with the identity I generate an algebra which is isomorphic to the $(2n+1)$ dimensional Heisenberg algebra. This algebra plays for the twisted convolution on $\mathbb{C}^n$ a role analogous to that of the Lie algebra of left invariant vector fields on a Lie group. In fact, it is easy to verify that

$$(1.3.10) \qquad Z_j(f \times g) = f \times Z_j g, \quad \overline{Z}_j(f \times g) = f \times \overline{Z}_j g$$

for $j = 1, 2, \ldots, n$. It is also easy to see that the following relations hold:

$$(1.3.11) \qquad W(Z_j f) = iW(f)A_j, \quad W(\overline{Z}_j f) = iW(f)A_j^*.$$

These are similar to the relations

$$(1.3.12) \qquad \left(\frac{\partial}{\partial x_j}f\right)\widehat{}(\xi) = i\xi_j \hat{f}(\xi)$$

satisfied by the Fourier transform.

The operator L is defined by

(1.3.13) $$L = -\tfrac{1}{2}\sum_{j=1}^{n}(Z_j\overline{Z}_j + \overline{Z}_j Z_j).$$

An easy calculation shows that L can be written in the form

(1.3.14) $$L = -\Delta_z + \tfrac{1}{4}|z|^2 - iN,$$

where N is the operator

(1.3.15) $$N = \sum_{j=1}^{n}\left(x_j\frac{\partial}{\partial x_j} - y_j\frac{\partial}{\partial x_j}\right).$$

We now prove that $\Phi_{\mu\nu}$ are eigenfunctions of the operator L.

Theorem 1.3.3. *One has the formulas*

(i) $$Z_j(\Phi_{\mu\nu}) = i(2\nu_j)^{\frac{1}{2}}\Phi_{\mu,\nu-e_j},$$

(ii) $$\overline{Z}_j(\Phi_{\mu\nu}) = i(2\nu_j+2)^{\frac{1}{2}}\Phi_{\mu,\nu+e_j}$$

for $j = 1, 2, \ldots, n$. Consequently, one has

(iii) $$L(\Phi_{\mu\nu}) = (2|\nu| + n)\Phi_{\mu\nu}.$$

Proof: As the functions $\Phi_{\mu\nu}(z)$ are the products of $\Phi_{\mu_j\nu_j}(z_j)$ it is enough to consider the case $n = 1$. So, we consider the functions

(1.3.16) $$\Phi_{j,k}(z) = (2\pi)^{-\frac{1}{2}}\int e^{ix\xi}h_j\left(\xi + \tfrac{1}{2}y\right)h_k\left(\xi - \tfrac{1}{2}y\right)d\xi.$$

Differentiating with respect to x and writing $2\xi = (\xi + \tfrac{1}{2}y) + (\xi - \tfrac{1}{2}y)$ we have

(1.3.17) $$\frac{\partial}{\partial x}\Phi_{jk}(z) = \frac{i}{2}(2\pi)^{-\frac{1}{2}}\left\{\int e^{ix\xi}(\xi+\tfrac{1}{2}y)h_j(\xi+\tfrac{1}{2}y)h_k(\xi-\tfrac{1}{2}y)d\xi \right. $$
$$\left. + \int e^{ix\xi}h_j(\xi+\tfrac{1}{2}y)(\xi-\tfrac{1}{2}y)h_k(\xi-\tfrac{1}{2}y)d\xi\right\}.$$

We also have

(1.3.18) $$i\frac{\partial}{\partial y}\Phi_{jk}(z) = \frac{i}{2}(2\pi)^{-\frac{1}{2}}\left\{\int e^{ix\xi}h'_j(\xi+\tfrac{1}{2}y)h_k(\xi-\tfrac{1}{2}y)d\xi \right.$$
$$\left. - \int e^{ix\xi}h_j(\xi+\tfrac{1}{2}y)h'_k(\xi-\tfrac{1}{2}y)d\xi\right\}.$$

Combining (1.3.17) and (1.3.18) and using the formulas

$$\left(-\frac{d}{dx}+x\right)h_k(x) = (2k+2)^{\frac{1}{2}}h_{k+1}(x)$$

and

$$\left(\frac{d}{dx}+x\right)h_k(x) = (2k)^{\frac{1}{2}}h_k(x)$$

we obtain

(1.3.19) $\quad \dfrac{\partial}{\partial z}\Phi_{jk}(z) = \dfrac{i}{2}\Big\{(2j+2)^{\frac{1}{2}}\Phi_{j+1,k}(z) + (2k)^{\frac{1}{2}}\Phi_{j,k-1}(z)\Big\},$

(1.3.20) $\quad \dfrac{\partial}{\partial \bar{z}}\Phi_{jk}(z) = \dfrac{i}{2}\Big\{(2j)^{\frac{1}{2}}\Phi_{j-1,k}(z) + (2k+2)^{\frac{1}{2}}\Phi_{j,k+1}(z)\Big\}.$

Writing $xe^{ix\xi} = -i\frac{\partial}{\partial \xi}e^{ix\xi}$ and integrating by parts we get

(1.3.21)
$$x\Phi_{jk}(z) = i(2\pi)^{-\frac{1}{2}}\Bigg\{\int e^{ix\xi}h'_j(\xi+\tfrac{1}{2}y)h_k(\xi-\tfrac{1}{2}y)d\xi$$
$$+ \int e^{ix\xi}h_j(\xi+\tfrac{1}{2}y)h'_k(\xi-\tfrac{1}{2}y)d\xi\Bigg\}.$$

We also have

(1.3.22)
$$iy\Phi_{jk}(z) = i(2\pi)^{-\frac{1}{2}}\Bigg\{\int e^{ix\xi}\Big((\xi+\tfrac{1}{2}y)-(\xi-\tfrac{1}{2}y)\Big)$$
$$\times h_j(\xi+\tfrac{1}{2}y)h_k(\xi-\tfrac{1}{2}y)d\xi\Bigg\}.$$

The last two formulas give

(1.3.23) $\quad (x+iy)\Phi_{jk}(z) = i\Big\{(2j)^{\frac{1}{2}}\Phi_{j-1,k}(z) - (2k+1)^{\frac{1}{2}}\Phi_{j,k+1}(z)\Big\}$

(1.3.24) $\quad (x-iy)\Phi_{jk}(z) = -i\Big\{(2j+2)^{\frac{1}{2}}\Phi_{j+1,k}(z) - (2k)^{\frac{1}{2}}\Phi_{j,k-1}(z)\Big\}.$

The formulas (i) and (ii) of the theorem now follow from the relations (1.3.19), (1.3.20), (1.3.23) and (1.3.24). The third one (iii) follows from (i), (ii) and the definition (1.3.13) of L. ∎

A similar calculation as above will show that $\Phi_{\mu\nu}$ are eigenfunctions of the Hermite operator $(-\Delta_z + \frac{1}{4}|z|^2)$ also. In fact, it can be proved that

(1.3.25) $$\left(-\Delta_z + \tfrac{1}{4}|z|^2\right)\Phi_{\mu\nu} = (|\nu| + |\mu| + n)\Phi_{\mu\nu}.$$

This justifies the name special Hermite functions, the name given by Strichartz [St2]. As we show below, the functions $\Phi_{\mu\nu}$ can also be expressed in terms of Laguerre polynomials. We first consider the functions $\Phi_{\mu\mu}$. Recall the definition of $L_k^\alpha(x)$ from Section 1.1. When $\alpha = 0$ we simply write $L_k(x)$.

Theorem 1.3.4.

$$\Phi_{\mu\mu}(z) = (2\pi)^{-\frac{n}{2}} \prod_{j=1}^{n} L_{\mu_j}\bigl(\tfrac{1}{2}|z_j|^2\bigr) e^{-\frac{1}{4}|z_j|^2}.$$

Proof: Again it is enough to consider the one dimensional case. We start with the Mehler's formula (see 1.1.11)

(1.3.26) $$\sum_{k=0}^{\infty} h_k\bigl(\xi + \tfrac{1}{2}y\bigr) h_k\bigl(\xi - \tfrac{1}{2}y\bigr) r^k$$
$$= \pi^{-\frac{1}{2}}(1-r^2)^{-\frac{1}{2}} e^{-\frac{1+r}{1-r}\frac{y^2}{4}} e^{-\frac{1-r}{1+r}\xi^2}.$$

If we take the Fourier transform in the ξ variable the left hand side becomes

(1.3.27) $$\sum_{k=0}^{\infty} \Phi_{k,k}(z) r^k.$$

The right hand side becomes

(1.3.28) $$\frac{1}{\sqrt{2\pi}}(1-r^2)^{-\frac{1}{2}} e^{-\frac{1+r}{1-r}\frac{y^2}{4}} \int e^{ix\xi} e^{-\frac{1-r}{1+r}\xi^2} d\xi$$
$$= \frac{1}{\sqrt{2\pi}}(1-r)^{-1} e^{-\frac{1}{4}\frac{1+r}{1-r}(x^2+y^2)}.$$

But this equals, by the generating function identity (1.1.33),

(1.3.29) $$\frac{1}{\sqrt{2\pi}} \sum_{k=0}^{\infty} L_k\bigl(\tfrac{1}{2}(x^2+y^2)\bigr) e^{-\frac{1}{4}(x^2+y^2)} r^k.$$

Hence we obtain

$$\Phi_{kk}(z) = (2\pi)^{-\frac{1}{2}} L_k\bigl(\tfrac{1}{2}|z|^2\bigr) e^{-\frac{1}{4}|z|^2}. \qquad\blacksquare$$

To state the next theorem we introduce some notation. If μ and m are multiindices we define

(1.3.30) $$L_\mu^m(z) = \prod_{j=1}^m L_{\mu_j}^{m_j}\left(\tfrac{1}{2}|z_j|^2\right).$$

We also write $\mu! = \mu_1!\mu_2!\cdots\mu_n!$ and $z^m = z_1^{m_1}\cdots z_n^{m_n}$. With these notations we have the following formulas.

Theorem 1.3.5.

(i) $\Phi_{\mu+m,\mu}(z) = (2\pi)^{-\frac{n}{2}}\left(\dfrac{\mu!}{(\mu+m)!}\right)^{\frac{1}{2}}\left(\dfrac{i}{\sqrt{2}}\right)^m \bar{z}^m L_\mu^m(z) e^{-\frac{1}{4}|z|^2},$

(ii) $\Phi_{\mu,\mu+m}(z) = (2\pi)^{-\frac{n}{2}}\left(\dfrac{\mu!}{(\mu+m)!}\right)^{\frac{1}{2}}\left(\dfrac{-i}{\sqrt{2}}\right)^m z^m L_\mu^m(z) e^{-\frac{1}{4}|z|^2}.$

Proof: From the definition it follows that $\overline{\Phi}_{\nu\mu}(-z) = \Phi_{\mu\nu}(z)$ and therefore it is enough to prove (i). Again we can assume that $n = 1$. So, we consider $\Phi_{j+1,j}(z)$ which can be written as

(1.3.31) $\Phi_{j+1,j}(z) = (-i)(2j+2)^{-\frac{1}{2}}\left(\dfrac{\partial}{\partial z} + \tfrac{1}{2}\bar{z}\right)\Phi_{j+1,j+1}(z)$

in view of Theorem 1.3.3. But now by the previous theorem

(1.3.32) $\Phi_{j+1,j+1}(z) = (2\pi)^{-\frac{1}{2}} L_{j+1}\left(\tfrac{1}{2}|z|^2\right) e^{-\frac{1}{4}|z|^2}$

and hence a calculation shows that

(1.3.33) $\left(\dfrac{\partial}{\partial z} + \tfrac{1}{2}\bar{z}\right)\Phi_{j+1,j+1}(z) = (2\pi)^{-\frac{1}{2}}\bar{z}(L_{j+1})'\left(\tfrac{1}{2}|z|^2\right) e^{-\frac{1}{4}|z|^2}.$

The derivative of the Laguerre polynomial satisfies the following relation (see 1.1.48)

(1.3.34) $$\dfrac{d}{dx} L_k^\alpha(x) = -L_{k-1}^{\alpha+1}(x),$$

and hence we have

(1.3.35) $\left(\dfrac{\partial}{\partial z} + \tfrac{1}{2}\bar{z}\right)\Phi_{j+1,j+1}(z) = (-1)(2\pi)^{-\frac{1}{2}}\bar{z} L_j^1\left(\tfrac{1}{2}|z|^2\right) e^{-\frac{1}{4}|z|^2}.$

This gives that

(1.3.36) $$\Phi_{j+1,j}(z) = \frac{i}{\sqrt{2}}(2\pi)^{-\frac{1}{2}}(j+1)^{-\frac{1}{2}}\bar{z}L_j^1(\tfrac{1}{2}|z|^2)e^{-\frac{1}{4}|z|^2}.$$

Now we can use an induction argument together with

(1.3.37) $$i(2j+2)^{\frac{1}{2}}\Phi_{j+m,j}(z) = \left(\frac{\partial}{\partial z} + \tfrac{1}{2}\bar{z}\right)\Phi_{j+m,j+1}(z)$$

to complete the proof. ∎

We now prove some interesting identities for the special Hermite functions. In the case of Euclidean convolution the equations $f * g = 0$ and $f * f = f$ do not admit nontrivial solutions. The next proposition shows that the twisted convolution equations $f \times g = 0$ and $f \times f = f$ do have nontrivial solution. This is another interesting feature of the twisted convolution which distinguishes it from the ordinary convolution.

Proposition 1.3.2.

(i) $\Phi_{\mu\nu} \times \Phi_{\alpha\beta} = 0$ if $\nu \neq \alpha$

(ii) $\Phi_{\mu\nu} \times \Phi_{\nu\beta} = (2\pi)^{\frac{n}{2}}\Phi_{\mu\beta}$.

Proof: We calculate the Weyl transform of $\overline{\Phi}_{\alpha\beta}$. For f and g in $L^2(\mathbb{R}^n)$

(1.3.38) $$(W(\overline{\Phi}_{\alpha\beta})f, g) = \int \overline{\Phi}_{\alpha\beta}(z)(W(z)f, g)dz$$
$$= (2\pi)^{\frac{n}{2}}(V(f,g), V(\Phi_\alpha, \Phi_\beta)).$$

In view of Proposition 1.3.1 we have

(1.3.39) $$(W(\overline{\Phi}_{\alpha\beta})f, g) = (2\pi)^{\frac{n}{2}}(f, \Phi_\alpha)(\Phi_\beta, g).$$

From this it follows that

(1.3.40) $$W(\overline{\Phi}_{\alpha\beta} \times \overline{\Phi}_{\mu\nu})f = (2\pi)^{\frac{n}{2}}(f, \Phi_\mu)W(\overline{\Phi}_{\alpha\beta})\Phi_\nu$$
$$= (2\pi)^{\frac{n}{2}}(f, \Phi_\mu)(\Phi_\nu, \Phi_\alpha)\Phi_\beta.$$

From this it is clear that $\overline{\Phi}_{\alpha\beta} \times \overline{\Phi}_{\mu\nu} = 0$ if $\alpha \neq \nu$ and $\overline{\Phi}_{\nu\beta} \times \overline{\Phi}_{\mu\nu} = (2\pi)^{\frac{n}{2}}\overline{\Phi}_{\mu\beta}$. The proposition follows from the fact that $\overline{f \times g} = \overline{g} \times \overline{f}$. ∎

We conclude this section with the following result which connects the Weyl transform, the Hermite projection operator P_k and the Laguerre function φ_k which is defined by

(1.3.41) $$\varphi_k(z) = L_k^{n-1}(\tfrac{1}{2}|z|^2)e^{-\frac{1}{4}|z|^2},$$

L_k^{n-1} being a Laguerre polynomial of type $(n-1)$.

Theorem 1.3.6.
$$W(\varphi_k) = (2\pi)^n P_k.$$

Proof: From the proof of the previous proposition we infer that
$$W(\Phi_{\mu\mu})f = (2\pi)^{\frac{n}{2}}(f,\Phi_\mu)\Phi_\mu.$$
As $P_k f = \sum_{|\mu|=k}(f,\Phi_\mu)\Phi_\mu$ it is enough to show that

(1.3.42) $$\varphi_k(z) = (2\pi)^{\frac{n}{2}} \sum_{|\mu|=k} \Phi_{\mu\mu}(z).$$

The Laguerre functions φ_k satisfy the generating function

(1.3.43) $$\sum_{k=0}^{\infty} \varphi_k(z) r^k = (1-r)^{-n} e^{-\frac{1}{2}\frac{1+r}{1-r}|z|^2}.$$

On the other hand, Theorem 1.3.4 gives

(1.3.44) $$\Phi_{\mu\mu}(z) = (2\pi)^{-\frac{n}{2}} \prod_{j=1}^{n} L_{\mu_j}\left(\tfrac{1}{2}|z_j|^2\right) e^{-\frac{1}{4}|z_j|^2}$$

and each $L_{\mu_j}(\tfrac{1}{2}|z_j|^2) e^{-\frac{1}{4}|z_j|^2}$ satisfy the relation

(1.3.45) $$\sum_{k=0}^{\infty} L_k\left(\tfrac{1}{2}|z_j|^2\right) e^{-\frac{1}{4}|z_j|^2} r^k = (1-r)^{-1} e^{-\frac{1}{2}\frac{1+r}{1-r}|z_j|^2}.$$

From (1.3.44) and (1.3.45) it is clear that

(1.3.46) $$\sum_{k=0}^{\infty} \Big(\sum_{|\mu|=k} \Phi_{\mu\mu}(z) \Big) r^k = (2\pi)^{-\frac{n}{2}} (1-r)^{-n} e^{-\frac{1}{2}\frac{1+r}{1-r}|z|^2}.$$

Comparing this with (1.3.43) we obtain (1.3.42). ∎

1.4 Hermite and Laguerre Functions on the Heisenberg Group

In this section we demonstrate how Hermite and Laguerre functions appear in the spectral theory of the sublaplacian $\mathcal{L}$ on the Heisenberg group.

Consider the reduced Heisenberg group H^n/Γ. Given a multiindex n, consider functions of the form

(1.4.1) $$f(z,t) = e^{it} e^{-im\cdot\theta} f_0(r_1,\ldots,r_n)$$

where $z_j = r_j e^{i\theta_j}$ and $m \cdot \theta = m_1\theta_1 + \cdots + m_n\theta_n$. Such functions are called $(-m)$-homogeneous. Let $L_m^p(H^n/\Gamma)$ stand for the subspace of $L^p(H^n/\Gamma)$ consisting of $(-m)$-homogeneous functions. The following proposition describes the spectrum of the sublaplacian $\mathcal{L}$ on the space $L_m^2(H^n/\Gamma)$.

Proposition 1.4.1. *The sublaplacian maps $L_m^2(H^n/\Gamma)$ into itself; on this space it has discrete spectrum. The normalized eigenfunctions are given by*
$$E_\mu(z,t) = (2\pi)^{-\frac{1}{2}} e^{it} \Phi_{\mu+m,\mu}(z),$$
with eigenvalue $(2|\mu| + n)$.

Proof: The sublaplacian $\mathcal{L}$ can be written in the form

(1.4.2) $\qquad \mathcal{L} = -\Delta_z - \frac{1}{4}|z|^2 \partial_t^2 - N\partial_t.$

From this it follows that if f is of the form (1.4.1), then

(1.4.3) $\qquad \mathcal{L}(f)(z,t) = e^{it} L(e^{-im\cdot\theta} f_0(r_1,\ldots,r_n)).$

To show that $\mathcal{L}f$ is of the same form as f we only need to show that $L(e^{-im\cdot\theta} f_0)$ is of the form $e^{-im\cdot\theta} g(r_1\ldots r_n)$. For the sake of simplicity we let $n = 1$. Introducing polar coordinates we have

(1.4.4) $\qquad \left(-\Delta + \frac{1}{4}|z|^2\right) = -\frac{\partial^2}{\partial r^2} - \frac{1}{r}\frac{\partial}{\partial r} - \frac{1}{r^2}\frac{\partial^2}{\partial \theta^2} + \frac{1}{4}r^2,$

and this gives

(1.4.5) $\qquad \left(-\Delta + \frac{1}{4}|z|^2\right)(e^{-im\theta} f_0(r)) = e^{-im\theta} g_0(r)$

with some g_0. On the other hand as f_0 is radial we have $N(f_0) = 0$ and a simple calculation shows that

(1.4.6) $\qquad N(e^{-im\theta}) = -ime^{im\theta},$

which gives $N(f_0(r)e^{-im\theta}) = -imf_0(r)e^{-im\theta}$. Hence $L(e^{-im\theta} f_0) = e^{-im\theta} g(r)$ for some g.

In view of (1.4.3) we see that

(1.4.7) $\qquad \mathcal{L}(E_\mu) = (2\pi)^{-\frac{1}{2}} e^{it} L(\Phi_{\mu+m,\mu}) = (2|\mu| + n) E_\mu$

since $\Phi_{\mu+m,\mu}$ are eigenfunctions of L. This proves the proposition. ∎

Thus we see that Laguerre functions appear as eigenfunctions of the sublaplacian on the reduced Heisenberg group. Using the above proposition one can obtain summability results for Laguerre series from corresponding results for the sublaplacian on the Heisenberg group.

For example, suppose for some kernel K_ϵ we have the almost everywhere convergence

$$(1.4.8) \qquad f(z,t) = \lim_{\epsilon \to 0} \sum_{k=0}^{\infty} K_\epsilon(2k+1)(f, E_k) E_k(z,t)$$

for f in $L^p(H^1/\Gamma)$. Since f and E_k are of the form (1.4.1) we obtain

$$(1.4.9) \qquad f_0(R) = \lim_{\epsilon \to 0} (2\pi)^{-\frac{1}{2}} \sum_{k=0}^{\infty} K_\epsilon(2k+1)(f, E_k) \mathcal{L}_k^m\left(\tfrac{1}{2} r^2\right)$$

where

$$(1.4.10) \qquad \mathcal{L}_k^m\left(\tfrac{1}{2} r^2\right) = \left(\tfrac{k!}{(k+m)!}\right)^{\frac{1}{2}} \left(\tfrac{1}{2} r^2\right)^{\frac{m}{2}} L_k^m\left(\tfrac{1}{2} r^2\right) e^{-\frac{1}{4} r^2}.$$

Now a calculation shows that

$$(1.4.11) \qquad (2\pi)^{-\frac{1}{2}} (f, E_k) = c(g, \mathcal{L}_k^m)$$

where $g(s) = f_0(\sqrt{2s})$ and c is a constant. Hence we have

$$(1.4.12) \qquad g(s) = \lim_{\epsilon \to 0} c \sum_{k=0}^{\infty} K_\epsilon(2k+1)(g, \mathcal{L}_k^m) \mathcal{L}_k^m(s).$$

The above arguments are taken from Dlugosz [Dl] where she has considered almost everywhere convergence of Laguerre series for the first time.

Next we investigate the relation between the Hermite expansions and the spectral decomposition of the sublaplacian. For the rest of the section we consider the polarized Heisenberg group $G = \mathbb{R}^n \times \mathbb{R}^n \times \mathbb{R}$ with the group law

$$(1.4.13) \qquad (x, y, t)(x', y', t') = (x+x', y+y', t+t'+x \cdot y').$$

We also consider a different sublaplacian

$$(1.4.14) \qquad \mathcal{L} = -\sum_{j=1}^{n} (P_j^2 + Q_j^2)$$

where $P_j = \frac{\partial}{\partial x_j}$ and $Q_j = \frac{\partial}{\partial y_j} + x_j \frac{\partial}{\partial t}$. The Schrödinger representation for the polarized Heisenberg group G corresponding to the parameter $\lambda = 1$ is given by

$$(1.4.15) \qquad \pi(x, y, t)\varphi(\xi) = e^{it} e^{iy \cdot \xi} \varphi(\xi + x).$$

The representation π defines a Lie algebra representation $d\pi$ by

(1.4.16) $$d\pi(X)f = \frac{d}{dt}\pi(\exp tX)f \Big|_{t=0}$$

where f is a smooth vector for π. Then one can easily verify that $d\pi(\mathcal{L}) = (-\Delta + |x|^2) = H$.

Let Γ_1 be the subgroup $\Gamma_1 = \{(x,y,t) : x, y \in \mathbb{Z}^n, t \in \mathbb{Z}\}$. Then G/Γ_1 is compact. Let $L_0^p(\mathbb{R}^n)$ denote the subspace of $L^p(\mathbb{R}^n)$ containing functions with compact support. We define a map $\tau : L_0^p(\mathbb{R}^n) \to L^p(G/\Gamma_1)$ by

(1.4.17) $$\tau f(x, y, t) = \sum_m e^{2\pi i t} e^{2\pi i m \cdot y} f(x + m)$$

where the sum is extended over all $m \in \mathbb{Z}^n$. Then τ extends as an isometry of $L^2(\mathbb{R}^n)$ into $L^2(G/\Gamma_1)$. If $1 \leq p < \infty$, then it can also be extended as a contraction of $L^p(\mathbb{R}^n)$ into $L^p(G/\Gamma_1)$. Let R be the quasi regular representation of G on $L^p(G/\Gamma_1)$. Then one has the relation

(1.4.18) $$\tau \pi(z, t) f = R(z, t) \tau f.$$

Using this we can write down a relation between the Hermite expansions and the eigenfunction expansions on the group G/Γ_1. The following arguments were used by Hulanicki and Jenkins [HJ] in proving a summability result for multiple Hermite expansions.

Let $e_\mu = \tau \Phi_\mu$ where Φ_μ are the Hermite functions on $\mathbb{R}^n$. Then it is easily checked that e_μ are eigenfunctions of the sublaplacian (1.4.14) with eigenvalue $(2|\mu| + n)$. Therefore, if we have a summability result for the eigenfunction expansions associated to the sublaplacian on G/Γ_1, then we can deduce a summability result for Hermite expansions. For example, as in the case of the reduced Heisenberg group, suppose that we have the almost everywhere convergence

(1.4.19) $$f(x, y, t) = \lim_{\varepsilon \to 0} \sum_\mu K_\varepsilon(2|\mu| + n)(f, e_\mu) e_\mu(x, y, t)$$

for f in $L^p(G/\Gamma_1)$. Let $f = \tau g$ with $g \in L^p(\mathbb{R}^n)$. Since $(f, e_\mu) = (g, \Phi_\mu)$ we have

(1.4.20) $$\sum_m e^{2\pi i m \cdot y} g(x + m)$$
$$= \lim_{\varepsilon \to 0} \sum_\mu K_\varepsilon(2|\mu| + n)(g, \Phi_\mu) \Big(\sum_m e^{2\pi i m \cdot y} \Phi_\mu(x + m) \Big)$$

for almost all $x, y \in Q$ where $Q = [0,1]^n$. Integrating out the y variable we obtain

(1.4.21) $$g(x) = \lim_{\varepsilon \to 0} \sum_{\mu} K_\varepsilon(2|\mu| + n)(g, \Phi_\mu)\Phi_\mu(x),$$

for almost every x. This proves our claim.

1.5 Asymptotic Properties of Hermite and Laguerre Functions

In the study of Hermite and Laguerre series we will make good use of certain pointwise and norm estimates for the Hermite and Laguerre functions. In this section we collect some pointwise estimates for them and briefly indicate how they can be used to get norm estimates. Detailed proofs can be found elsewhere. First we have the following estimates for the Hermite functions. For a proof we refer to [GS].

Lemma 1.5.1. Let $N = (2n + 1)$. The Hermite functions satisfy the estimates

$$|h_n(x)| \leq C(N^{\frac{1}{3}} + |x^2 - N|)^{-\frac{1}{4}}, \quad x^2 \leq 2N$$
$$\leq Ce^{-\gamma x^2}, \quad x^2 > 2N$$

where γ is a fixed positive constant. We also have the estimate

$$|h_n(x)| \leq CN^{-\frac{1}{6}}(x - N^{\frac{1}{2}})^{-\frac{1}{4}} e^{-\varepsilon N^{\frac{1}{4}}(x - N^{\frac{1}{2}})^{\frac{3}{2}}}$$

for $N^{\frac{1}{2}} + N^{-\frac{1}{6}} \leq x \leq (2N)^{\frac{1}{2}}$ where $\varepsilon > 0$ is a fixed constant.

These estimates are good enough to get upper bounds for the L^p norms of the Hermite functions. But when we want to prove lower bounds more precise asymptotic properties have to be used. In particular, one requires the following asymptotic property:

(1.5.1) $$h_n(x) = \left(\frac{2}{\pi}\right)^{\frac{1}{2}} (N - x^2)^{-\frac{1}{4}} \cos\left(\frac{N(2\theta - \sin\theta) - \pi}{4}\right)$$
$$+ O(N^{\frac{1}{2}}(N - x^2)^{-\frac{7}{4}})$$

where $0 \leq x \leq N^{\frac{1}{2}} - N^{-\frac{1}{6}}$ and $\theta = \cos^{-1}(xN^{-\frac{1}{2}})$. We can now state the norm estimates for the Hermite functions.

HERMITE AND LAGUERRE FUNCTIONS

Lemma 1.5.2. *As $n \to \infty$ the Hermite functions satisfy the estimates*

(i) $\|h_n\|_p \sim n^{\frac{1}{2p}-\frac{1}{4}}$, $\quad 1 \leq p < 4$,

(ii) $\|h_n\|_p \sim n^{-\frac{1}{8}} \log n$, $\quad p = 4$,

(iii) $\|h_n\|_p \sim n^{-\frac{1}{6p}-\frac{1}{12}}$, $\quad 4 < p \leq \infty$.

Here $a_n \sim b_n$ means $a_n = O(b_n)$ and $b_n = O(a_n)$.

Proof: Since $h_n(x)$ is either even or odd, one has

$$\|h_n\|_p = 2^{\frac{1}{p}} \left(\int_0^\infty |h_n(x)|^p dx \right)^{\frac{1}{p}}.$$

We can split the integral into three parts corresponding to the intervals $0 \leq x \leq \sqrt{N}$, $\sqrt{N} \leq x \leq \sqrt{2N}$ and $x \geq \sqrt{2N}$. Using Lemma 1.5.1 we obtain three upper bounds, and comparing them for various values of p we can establish the upper bounds of the Lemma. In order to prove the lower bounds we can consider the interval $1 \leq x \leq N^{\frac{1}{2}} - N^{-\frac{1}{6}}$, use the estimate (1.5.1) together with Lemma 15 of Muckenhoupt [Mu2]. The details are omitted. ∎

We now state some asymptotic properties of the normalized Laguerre functions $\mathcal{L}_n^\alpha(x)$. We let $\nu = 4n + 2\alpha + 2$ and assume $\alpha > -1$.

Lemma 1.5.3. *The Laguerre functions satisfy*

$$|\mathcal{L}_n^\alpha(x)| \leq C \begin{cases} (x\nu)^{\alpha/2}, & 0 \leq x \leq \frac{1}{\nu} \\ (x\nu)^{-\frac{1}{4}}, & \frac{1}{\nu} \leq x \leq \nu/2 \\ \nu^{-\frac{1}{4}}(\nu^{\frac{1}{3}} + |\nu - x|)^{-\frac{1}{4}}, & \frac{\nu}{2} \leq x \leq \frac{3\nu}{2} \\ e^{-\gamma x}, & x \geq \frac{3\nu}{2} \end{cases}$$

where $\gamma > 0$ is a fixed constant.

As in the case of Hermite functions we have the following asymptotic property of the Laguerre function $\mathcal{L}_n^\alpha(x)$ in the interval $1 \leq x \leq \nu - \nu^{\frac{1}{3}}$.

$$(1.5.2) \quad \mathcal{L}_n^\alpha(x) = \frac{(2/\pi)^{\frac{1}{2}}(-1)^n}{x^{\frac{1}{4}}(\nu-x)^{\frac{1}{4}}} \cos g + O\left(\frac{1}{(\nu x)^{\frac{3}{4}}} + \frac{\nu^{\frac{1}{4}}}{(\nu-x)^{\frac{7}{4}}} \right)$$

where $g = (\nu(2\theta - \sin 2\theta) - \pi)/4$. Again this is needed only in getting a lower bound for the L^p norms of the Laguerre functions.

Lemma 1.5.4. Let $\alpha + \beta > -1$ and $\alpha > -2/p$, $1 \leq p < \infty$. Then as $n \to \infty$ the following estimates are valid.

(i) $\|\mathcal{L}_n^{\alpha+\beta}(x)x^{-\beta/2}\|_p \sim n^{\frac{1}{p}-\frac{1}{2}-\frac{\beta}{2}}$, if $\beta < \frac{2}{p} - \frac{1}{2}$

(ii) $\|\mathcal{L}_n^{\alpha+\beta}(x)x^{-\beta/2}\|_p \sim n^{\frac{\beta}{2}-\frac{1}{p}}$, if $\beta > \frac{2}{p} - \frac{1}{2}$

for p in the interval $1 \leq p \leq 4$; when $p > 4$

(iii) $\|\mathcal{L}_n^{\alpha+\beta}(x)x^{-\beta/2}\|_p \sim n^{-\frac{1}{3}+\frac{1}{3p}-\frac{\beta}{2}}$, if $\beta \leq \frac{4}{3p} - \frac{1}{3}$

(iv) $\|\mathcal{L}_n^{\alpha+\beta}(x)x^{-\beta/2}\|_p \sim n^{\frac{\beta}{2}-\frac{1}{p}}$, if $\beta > \frac{4}{3p} - \frac{1}{3}$.

Proof: As in the case of Lemma 1.5.2, we estimate the integrand in various intervals using the estimates of Lemma 1.5.3. A comparison of the bounds occurring on the different subintervals yields the upper estimates of the lemma. The lower estimate is obtained by establishing three different lower bounds and choosing the largest one among them for each β. One needs to use (1.5.2) and Lemma 15 of [Mu2]. For more details we refer the reader to [Ma2]. ∎

In Chapter 7, while proving the transplantation theorem we need several other formulas satisfied by the Laguerre functions. We also need some estimates for the Bessel functions $J_\alpha(z)$ and $J_\alpha(iz)$. As we are not going to prove any of those formulas, it may be better to state them wherever they are needed. That will save the reader from the painful job of going back and forth.

CHAPTER 2
SPECIAL HERMITE EXPANSIONS

In this chapter we study the L^p mapping properties of the Riesz means and multipliers for special Hermite expansions. A multiplier theorem is proved using Littlewood–Paley–Stein theory of g functions defined by the special Hermite semigroup e^{-tL} generated by the operator L. The g functions, in turn, are studied using the theory of oscillatory singular integrals a special kind of which is also developed here. The oscillatory singular integrals are also applied to study Riesz transforms for the operator L. The Riesz means are studied using the method of restriction theorems. Almost everywhere and mean convergence of the Riesz means are proved.

2.1 Special Hermite Expansions

Given a function f in $L^p(\mathbb{C}^n)$, $1 \leq p \leq \infty$ we can formally expand f in terms of the special Hermite functions as

(2.1.1) $$f(z) = \sum_\mu \sum_\nu (f, \Phi_{\mu\nu}) \Phi_{\mu\nu}(z).$$

As $\Phi_{\mu\nu}$ are Schwartz class functions the coefficients

(2.1.2) $$(f, \Phi_{\mu\nu}) = \int_{\mathbb{C}^n} f(z) \overline{\Phi}_{\mu\nu}(z) dz$$

are well defined. For f in $L^2(\mathbb{C}^n)$ the above series converges to f in the L^2 norm. We are interested in the convergence of the series when f is in $L^p(\mathbb{C}^n)$, $1 \leq p \leq \infty$. We claim that the above series can be put in the compact form

(2.1.3) $$f(z) = (2\pi)^{-\frac{n}{2}} \sum_\nu f \times \Phi_{\nu\nu}(z).$$

This can be seen as follows. If f has the expansion (2.1.1) then

(2.1.4) $$f \times \Phi_{\alpha\alpha} = \sum_\mu \sum_\nu (f, \Phi_{\mu\nu}) \Phi_{\mu\nu} \times \Phi_{\alpha\alpha}.$$

In view of Proposition 1.3.2 the above sum reduces to

(2.1.5) $$f \times \Phi_{\alpha\alpha} = (2\pi)^{\frac{n}{2}} \sum_{\mu} (f, \Phi_{\mu\alpha}) \Phi_{\mu\alpha}.$$

This proves the claim.

In (2.1.3) we can first sum over all ν with $|\nu| = k$ and noting that

$$\sum_{|\nu|=k} \Phi_{\nu\nu}(z) = (2\pi)^{-\frac{n}{2}} \varphi_k(z)$$

we can also write the special Hermite expansion in the form

(2.1.6) $$f(z) = (2\pi)^{-n} \sum_{k=0}^{\infty} f \times \varphi_k(z).$$

It is clear that $f \times \varphi_{\nu\nu}$ is an eigenfunction of the operator L with eigenvalue $(2|\nu| + n)$. Hence $(2\pi)^{-n} f \times \varphi_k$ is the projection of f onto the eigenspace corresponding to the eigenvalue $(2k + n)$. Note that unlike the Hermite case the eigenspaces are infinite dimensional here. Since (2.1.6) is the eigenfunction expansion associated to the elliptic operator L, it follows from a transplantation theorem of Kenig–Stanton–Tomas [KST] that the partial sums will not converge for functions in $L^p(\mathbb{C}^n)$, $p \neq 2$ in the norm. In the next chapter we apply this transplantation theorem to the Hermite operator H. Here we will not go into the details of the transplantation.

As the series (2.1.6) fail to converge we are led to consider the Cesàro means

(2.1.7) $$\sigma_N^\delta f(z) = \frac{(2\pi)^{-n}}{A_N^\delta} \sum_{k=0}^{N} A_{N-k}^\delta f \times \varphi_k(z)$$

where A_k^δ are the binomial coefficients. We can also consider the Riesz means

(2.1.8) $$S_R^\delta f(z) = (2\pi)^{-n} \sum \left(1 - \frac{2k+n}{R}\right)_+^\delta f \times \varphi_k(z).$$

The behavior of these two means are the same in the sense that for each δ, $S_R^\delta f$ converges to f in the norm if and only if the same is true of $\sigma_N^\delta f$. In Section 2.5 we study the convergence properties of S_R^δ. We also make use of some properties of the Cesàro means. Using another version of the transplantation theorem it can be proved that for functions f in $L^1(\mathbb{C}^n)$,

SPECIAL HERMITE EXPANSIONS   31

$S_R^\delta f$ will not converge in the norm unless $\delta > n - \frac{1}{2}$. This means that $n - \frac{1}{2}$ is the smallest index with the property that $\delta > n - \frac{1}{2}$ implies L^1 convergence of $S_R^\delta f$. $(n - \frac{1}{2})$ is therefore the critical index for the Riesz (or Cesàro) summability of the special Hermite expansions. That $(n-\frac{1}{2})$ is the critical index can also be proved directly without recourse to the transplantation theorem by considering radial functions and using some estimates for Laguerre functions.

In Section 2.4 we consider certain multiplier transforms for the special Hermite expansions. These operators are formally defined by

$$(2.1.9) \qquad T_m f(z) = (2\pi)^{-\frac{n}{2}} \sum_\nu m(\nu) f \times \Phi_{\nu\nu}(z)$$

where m is a bounded function defined on the set of all multiindices. This operator is clearly bounded on $L^2(\mathbb{C}^n)$ but may not be bounded on $L^p(\mathbb{C}^n)$, $p \neq 2$ unless some more conditions are imposed on the function m. In Theorem 2.4.1 we give a sufficient condition on m which ensures that T_m is bounded on $L^p(\mathbb{C}^n)$, $1 < p < \infty$.

The above operator T_m can also be interpreted as a multiplier for the Weyl transform. Taking the Weyl transform of $T_m f$ and noting that $W(\Phi_{\nu\nu})$ is the projection operator

$$(2.1.10) \qquad W(\Phi_{\nu\nu})g = (2\pi)^{\frac{n}{2}}(g, \Phi_\nu)\Phi_\nu, \quad g \in L^2(\mathbb{R}^n)$$

we get the relation

$$(2.1.11) \qquad W(T_m f) = W(f)m(A)$$

where $m(A)$ is the bounded operator acting on $L^2(\mathbb{R}^n)$ given by

$$(2.1.12) \qquad m(A)g = \sum_\nu m(\nu)(g, \Phi_\nu)\Phi_\nu.$$

More generally, given an operator M bounded on $L^2(\mathbb{R}^n)$, we can associate another operator T_M on $L^2(\mathbb{C}^n)$ by the prescription

$$(2.1.13) \qquad W(T_M f) = W(f)M.$$

If T_M extends to a bounded operator on $L^p(\mathbb{C}^n)$ we say that M is an L^p multiplier for the Weyl transform. Thus the boundedness of T_m means that $m(A)$ is an L^p multiplier for the Weyl transform. General Weyl multipliers have been studied in [M].

Multipliers of some special form, which don't fall into the general class of multipliers studied in Section 2.4 are also important. One such example is considered in the next section, called the Riesz transform for the special Hermite expansions. These are the operators defined by

(2.1.14) $$R_j f = Z_j L^{-\frac{1}{2}} f, \; j = 1, 2, \ldots, n$$

and correspond to the multipliers $m_j(A) = H^{-\frac{1}{2}} A_j$. These operators are studied using oscillatory singular integrals. We conclude this section with the following remark concerning a transference result. If M is an L^p multiplier for the Weyl transform, then Mauceri has proved (see [M]) that M is a bounded operator on $L^p(\mathbb{R}^n)$. In view of this we can deduce certain results for the Hermite expansions from the corresponding results for the special Hermite expansions. Such results are not optimal and so we do not follow that path.

2.2 Oscillatory Singular Integrals and Riesz Transforms

Oscillatory singular integral operators have been studied by Ricci-Stein [RS] and Chanillo-Christ [CC]. These are operators of the form

(2.2.1) $$Tf(x) = p.v. \int_{\mathbb{R}^n} e^{iP(x,y)} K(x-y) f(y) dy$$

where K is a standard Calderon-Zygmund kernel in $\mathbb{R}^n$ and P is a general real valued polynomial in x and y. Such operators have arisen in connection with singular integrals on lower dimensional varieties, on the Heisenberg group in relation to twisted convolution and also as model operators occurring in the theory of singular Radon tranforms. In this section we are concerned with the simplest case where

(2.2.2) $$Tf(z) = p.v. \int_{\mathbb{C}^n} e^{i/2 Im z \cdot \bar{w}} K(z-w) f(w) dw.$$

Here K is assumed to be a Calderon-Zygmund kernel on $\mathbb{C}^n$. Thus K satisfies the estimates

(2.2.3) $$|K(z)| \le C|z|^{-2n},$$

(2.2.4) $$|\nabla K(z)| \le C|z|^{-2n-1}.$$

For operators of the form (2.2.2) we prove the following theorem.

Theorem 2.2.1. *Assume that the operator T defined in (2.2.2) is bounded on $L^2(\mathbb{C}^n)$. Then for $1 < p < \infty$, it is also bounded on $L^p(\mathbb{C}^n)$.*

In proving this theorem we closely follow [RS]. We choose a cutoff function α in $C_0^\infty(\mathbb{C}^n)$ with the property that α is supported in $|z| \leq 1$ and $\alpha = 1$ for $|z| \leq \frac{3}{4}$. We decompose T into two parts

(2.2.5) $$Tf = T_0 f + T_\infty f$$

where T_0 (resp, T_∞) is $K_0 \times f$ (resp., $K_\infty \times f$) where $K_0(z) = K(z)\alpha(z)$ (resp., $K_\infty(z) = K(z)(1 - \alpha(z))$). We first consider the local part. We proceed step by step in proving that T_0 is bounded on $L^p(\mathbb{C}^n)$.

Lemma 2.2.1. *T_0 is bounded on $L^2(\mathbb{C}^n)$.*

Proof: We split f as the sum of three functions, $f = f_1 + f_2 + f_3$ where

(2.2.6) $$f_1(z) = f(z)\mathcal{X}\left(|z - \zeta| \leq \tfrac{1}{2}\right),$$

(2.2.7) $$f_2(z) = f(z)\mathcal{X}\left(\tfrac{1}{2} < |z - \zeta| \leq \tfrac{5}{4}\right)$$

and $f_3 = f - f_1 - f_2$ where $\zeta \in \mathbb{C}^n$ is fixed. We claim that

(2.2.8) $$\int_{|z-\zeta|\leq \frac{1}{4}} |T_0 f(z)|^2 dz \leq C \int_{|w-\zeta|\leq \frac{5}{4}} |f(w)|^2 dw.$$

Once we have this, then an integration with respect to ζ will prove the lemma.

We first consider $T_0 f_1$. Since we are interested in $|z - \zeta| \leq \frac{1}{4}$ and on the support of f_1 we have $|w - \zeta| \leq \frac{1}{2}$ we see that $|z - w| \leq \frac{3}{4}$. By the choice of α it follows that

$$T_0 f_1 = \int e^{i/2 Im z \overline{w}} K(z - w) \alpha(z - w) f_1(w) dw$$

$$= T f_1(z)$$

whenever $|z - \zeta| \leq \frac{1}{4}$. This proves (2.2.8) with f_1 in place of f. In the case of f_2, $\frac{1}{2} < |w - \zeta| \leq \frac{5}{4}$ and so when $|z - \zeta| \leq \frac{1}{4}$ we have $\frac{1}{4} < |w - z| \leq \frac{3}{2}$. As the kernel is integrable over this region we obtain (2.2.8) for f_2 also. Finally, as f_3 is supported in $|w - \zeta| > \frac{5}{4}$ and α in $|z - w| \leq 1$ we get $T_0 f_3 = 0$.

This completes the proof of (2.2.8). ■

Lemma 2.2.2. *The operator $\widetilde{T}_0$ defined by*

$$\widetilde{T}_0 f(z) = \int_{\mathbb{C}^n} K(z-w)\alpha(z-w)f(w)dw$$

is bounded in $L^p(\mathbb{C}^n)$ for $1 < p < \infty$.

Proof: Since $\widetilde{T}_0$ is a Calderon–Zygmund operator it is enough to show that it is bounded on $L^2(\mathbb{C}^n)$. We write

$$(2.2.9) \quad \widetilde{T}_0 f(z) = \int_{\mathbb{C}^n} K_0(z-w)e^{i/2\mathrm{Im}z\overline{w}}f(w)dw$$
$$+ \int_{\mathbb{C}^n} (1 - e^{i/2\mathrm{Im}z\overline{w}})K_0(z-w)f(w)dw$$
$$= T_0 f(z) + \int_{\mathbb{C}^n} \widetilde{K}(z,w)f(w)dw.$$

As before we will show that

$$(2.2.10) \quad \int_{|z-\zeta|\leq 1} |\widetilde{T}_0 f(z)|^2 dz \leq C \int_{|w-\zeta|\leq 2} |f(w)|^2 dw$$

which will prove the lemma.

This time we split the function f into two parts, $f = f_1 + f_2$ with $f_1(z) = f(z)\mathcal{X}(|z-\zeta| \leq 2)$. Since the kernel of $\widetilde{T}_0$ is supported in $|z - w| \leq 1$ it follows that $\widetilde{T}_0 f_2(z) = 0$ for $|z - \zeta| \leq 1$. Thus

$$(2.2.11) \quad \widetilde{T}_0 f(z) = T_0 f_1(z) + \int_{\mathbb{C}^n} \widetilde{K}(z,w)f_1(w)dw.$$

By mean value theorem, the kernel $\widetilde{K}$ satisfies the estimate

$$(2.2.12) \quad |\widetilde{K}(z,w)| \leq C|z-w|^{-2n+1}$$

so that it is integrable. By this observation and Lemma 2.2.1 it follows that (2.2.10) is true. ∎

Lemma 2.2.3. T_0 *is bounded on $L^p(\mathbb{C}^n)$.*

Proof: We use the same trick as before. We claim, for each $\zeta \in \mathbb{C}^n$,

$$(2.2.13) \quad \int_{|z|\leq 1} |T_0 f(z+\zeta)|^p dz \leq C \int_{|w|\leq 2} |f(w+\zeta)|^p dw.$$

To this end we first observe that

$$\operatorname{Im}(z+\zeta)\cdot(\overline{w}+\overline{\zeta}) = \operatorname{Im} z\cdot\overline{\zeta} + \operatorname{Im} z\cdot\overline{w} + \operatorname{Im}\zeta\cdot\overline{w}$$

so that making a change of variables

(2.2.14) $\quad T_0 f(z+\zeta)$
$$= e^{i/2\operatorname{Im} z\cdot\overline{\zeta}} \int_{\mathbb{C}^n} K_0(z-w) e^{i/2\operatorname{Im} z\overline{w}} f(\zeta+w) e^{i/2\operatorname{Im}\zeta\cdot\overline{w}} dw.$$

As before, we write this as

(2.2.15) $\quad T_0 f(z+\zeta) e^{-i/2\operatorname{Im} z\cdot\overline{\zeta}} = \widetilde{T}_0 g(z) + \int_{\mathbb{C}^n} \widetilde{K}(z,w) g(w) dw$

where $g(w) = f(\zeta+w) e^{i/2\operatorname{Im}\zeta\cdot\overline{w}}$. For $|z| \leq 1$ only the points with $|w| \leq 2$ matter and hence the kernel $\widetilde{K}$ is integrable. By the previous lemma it therefore follows that (2.2.13) is valid. An integration with respect to ζ now proves the Lemma. ∎

Thus we have taken care of the local part T_0 in (2.2.5). To deal with T_∞ we decompose the kernel K_∞ using a partition of unity. Let $\psi \in C_0^\infty(\frac{1}{2} \leq |z| \leq 2)$ be a nonnegative function with the property that $\psi(z) = 1$ for $\frac{1}{4} \leq |z| \leq \frac{5}{4}$ and $\sum_{-\infty}^{\infty} \psi(2^{-j}z) = 1$. Let us write $K_j(z) = K_\infty(z)\psi(2^{-j}z)$ so that

(2.2.16) $$K_\infty(z) = \sum_{j=0}^{\infty} K_j(z).$$

The kernel $K_j(z)$ is supported in $2^{j-1} \leq |z| \leq 2^{j+1}$ and satisfies $|K_j(z)| \leq C|z|^{-2n}$. Let $T_j f = K_j \times f$ and T_j^* be the adjoint of T_j acting on $L^2(\mathbb{C}^n)$.

In the following lemma we get an estimate for the operator norm of $T_j^* T_j$. In the general case of operators of the form (2.2.1) a proof is given in [RS]. The proof was much more difficult in that generality. Here we present a simple proof in our situation, thanks to the fact that twisted convolution with an L^2 function is a bounded operator on $L^2(\mathbb{C}^n)$.

Lemma 2.2.4. *There is an $\varepsilon > 0$ such that the operator norm $\|T_j\|_p$ of T_j on $L^p(\mathbb{C}^n)$, $1 < p < \infty$ satisfies $\|T_j\|_p \leq C 2^{-\varepsilon j}$.*

Proof: We show that $\|T_j\|_2 \leq C 2^{-nj}$. As the kernels K_j are uniformly integrable $\|T_j\|_1 \leq C$ follows. Interpolating these two estimates we obtain the lemma. To get the estimate for $\|T_j\|_2$ we estimate $\|T_j^* T_j\|_2$. To this end we calculate the kernel of $T_j^* T_j$.

Let $K_j^*(z,w)$ be the kernel of the adjoint T_j^*. Then an easy calculation shows that

$$(2.2.17) \qquad K_j^*(z,w) = \overline{K_j}(w-z)e^{i/2\mathrm{Im}z\cdot\overline{w}}.$$

The kernel $L_j(z,w)$ of $T_j^*T_j$ is therefore given by

$$(2.2.18) \qquad L_j(z,w) = \int_{\mathbb{C}^n} \overline{K_j}(\zeta-z)e^{i/2\mathrm{Im}z\cdot\overline{\zeta}} K_j(\zeta-w) e^{i/2\mathrm{Im}\zeta\cdot\overline{w}} d\zeta.$$

Making a change of variables we easily calculate that

$$(2.2.19) \qquad L_j(z,w) = G_j(z-w) e^{i/2\mathrm{Im}z\cdot\overline{w}}$$

where the function $G_j(z)$ is given by

$$(2.2.20) \qquad G_j(z) = \int_{\mathbb{C}^n} \overline{K_j}(w) K_j(z+w) e^{i/2\mathrm{Im}z\cdot\overline{w}} dw.$$

Thus we have proved that $T_j^* T_j f = G_j \times f$.

By Theorem 1.2.2 it follows that

$$(2.2.21) \qquad \|T_j^* T_j f\|_2 \le C \|G_j\|_2 \|f\|_2.$$

As G_j is defined as the twisted convolution of K_j with $\overline{K_j}$ it is clear that

$$(2.2.22) \qquad \|G_j\|_2 \le C \|K_j\|_2^2.$$

Since K_j is supported in $2^{j-1} \le |z| \le 2^{j+1}$ and satisfies $|K_j(z)| \le C|z|^{-2n}$ it is immediate that

$$(2.2.23) \qquad \|K_j\|_2 \le C \int_{2^{j-1}}^{2^{j+1}} r^{-2n-1} dr \le C 2^{-2nj}.$$

Hence $\|T_j^* T_j\|_2 \le C 2^{-4nj}$ which proves that $\|T_j\|_2 \le C 2^{-2nj}$. This proves the Lemma. ∎

Thus we have proved that the series $\sum_{j=0}^{\infty} T_j$ converges in the operator norm and consequently T_∞ is bounded on $L^p(\mathbb{C}^n)$, $1 < p < \infty$. This completes the proof of Theorem 2.2.1.

We now study the mapping properties of the operators R_j and $\overline{R}_j$ defined by

(2.2.24) $\qquad R_j f = Z_j L^{-\frac{1}{2}} f, \quad \overline{R}_j f = \overline{Z}_j L^{-\frac{1}{2}} f, \quad j = 1, 2, \ldots, n.$

In view of Theorem 1.3.3 these operators can be put in the form

(2.2.25) $\qquad R_j f = i(2\pi)^{-\frac{n}{2}} \sum_\mu \left(\dfrac{2\mu_j}{2|\mu|+n}\right)^{\frac{1}{2}} f \times \Phi_{\mu,\mu-e_j},$

(2.2.25') $\qquad \overline{R}_j f = i(2\pi)^{-\frac{n}{2}} \sum_\mu \left(\dfrac{2\mu_j+2}{2|\mu|+n}\right)^{\frac{1}{2}} f \times \Phi_{\mu,\mu+e_j}.$

From this form it is clear that R_j and $\overline{R}_j$ are bounded on $L^2(\mathbb{C}^n)$.

Theorem 2.2.2. *The Riesz transforms R_j and $\overline{R}_j$, $j = 1, 2, \ldots, n$ are bounded on $L^2(\mathbb{C}^n)$, $1 < p < \infty$, and also of weak type $(1, 1)$.*

Proof: The idea of the proof is to show that R_j and $\overline{R}_j$ are twisted convolution operators with Calderon–Zygmund kernels. So, we can appeal to Theorem 2.2.1 to prove that they are bounded on $L^p(\mathbb{C}^n)$, $1 < p < \infty$. The weak type $(1, 1)$ inequality will follow from a theorem of Chanillo–Christ [CC] which we have not proved here.

The operator $L^{-\frac{1}{2}}$ is defined using spectral theorem. If e^{-tL} is the semigroup generated by the operator L then $L^{-\frac{1}{2}}$ can be represented as

(2.2.26) $\qquad L^{-\frac{1}{2}} f = \dfrac{1}{\sqrt{\pi}} \int_0^\infty t^{-\frac{1}{2}} e^{-tL} f \, dt.$

For f in $L^2(\mathbb{C}^n)$, the function $e^{-tL} f$ has the special Hermite expansion

(2.2.27) $\qquad e^{-tL} f = (2\pi)^{-n} \sum_{k=0}^\infty e^{-(2k+n)t} f \times \varphi_k.$

As such $e^{-tL} f$ is given by twisted convolution with the kernel

(2.2.28) $\qquad K_t(z) = (2\pi)^{-n} \sum_{k=0}^\infty e^{-(2k+n)t} \varphi_k(z).$

Now, in view of the generating function (1.1.45) we can easily calculate the kernel to be

(2.2.29) $\qquad K_t(z) = (4\pi)^{-n} (\sinh t)^{-n} e^{-\frac{1}{4}|z|^2 (\coth t)}.$

Therefore, it follows that

(2.2.30) $$L^{-\frac{1}{2}}f = \frac{1}{\sqrt{\pi}} \int_0^\infty t^{-\frac{1}{2}} f \times K_t \, dt.$$

From this we conclude, using (1.3.10), that

(2.2.31) $$R_j f = \frac{1}{\sqrt{\pi}} \int_0^\infty t^{-\frac{1}{2}} (f \times Z_j K_t) dt$$

and a similar expression holds for $\overline{R}_j$ also.

The proof of the theorem will be complete once we show that the kernels defined by

(2.2.32) $$K_j(z) = \frac{1}{\sqrt{\pi}} \int_0^\infty t^{-\frac{1}{2}} Z_j K_t(z) dt$$

are Calderon–Zygmund kernels. That is the content of the next lemma.

Lemma 2.2.5. *The kernels K_j satisfy*

(i) $|K_j(z)| \leq C|z|^{-2n}$,

(ii) $|\nabla K_j(z)| \leq C|z|^{-2n-1}$.

Proof: Up to a constant multiple the kernel K_j is given by the integral

(2.2.33) $$\left(\frac{\partial}{\partial z_j} + \frac{1}{2}\bar{z}_j\right) \int_0^\infty t^{-\frac{1}{2}} (\sinh t)^{-n} e^{-\frac{1}{4}|z|^2 (\coth t)} dt.$$

Taking Z_j under the integral sign the above equals

(2.2.34) $$-\tfrac{1}{4}\bar{z}_j \int_0^\infty t^{-\frac{1}{2}} e^{-t} (\sinh t)^{-n-1} e^{-\frac{1}{4}|z|^2 (\coth t)}.$$

The part of the integral taken from one to infinity has an exponential decay in $|z|$. When $0 < t < 1$, $\sinh t = O(t)$ and $\coth t = O(t^{-1})$ and consequently

(2.2.35) $$\left| \int_0^1 t^{-\frac{1}{2}} e^{-t} (\sinh t)^{-n-1} e^{-\frac{1}{4}|z|^2 \coth t} dt \right|$$
$$\leq C \int_0^1 t^{-n-\frac{3}{2}} e^{-\frac{a}{t}|z|^2} dt \leq C|z|^{-2n-1}$$

This proves (i). Part (ii) is proved in a similar way. ■

Thus we have proved that the Riesz transform are bounded on $L^p(\mathbb{C}^n)$, $1 < p < \infty$. Since R_j and $\overline{R}_j$ are multipliers for the Weyl transform, the multipliers being $A_j H^{-\frac{1}{2}}$ and $A_j^* H^{-\frac{1}{2}}$ it follows by the transference result mentioned in Section 2.2 that $A_j H^{-\frac{1}{2}}$ and $A_j^* H^{-\frac{1}{2}}$ are bounded on $L^p(\mathbb{R}^n)$, $1 < p < \infty$. We will give an independent proof of this result in Chapter 4 without recourse to transference.

2.3 Littlewood–Paley–Stein Theory

In the next section we plan to study the L^p mapping properties of multiplier operators T_m defined in (2.1.9). In the previous section we were fortunate in the study of the Riesz transforms as they turned out to be oscillatory singular integrals. But the operators T_m are not so easily accessible. Oscillatory singular integrals do enter into the picture but in an indirect way. To study T_m we develop a Littlewood–Paley–Stein theory for the semigroup e^{-tL}. It is in developing this theory we need a vector valued analogue of Theorem 2.2.1.

A Littlewood–Paley–Stein theory of g functions can be developed in the more general context of symmetric diffusion semigroups. In this section we consider the semigroup $T^t = e^{-tL}$ defined in (2.2.27). For each integer $k \geq 1$ we define

$$(2.3.1) \qquad (g_k(f,z))^2 = \int_0^\infty |\partial_t^k T^t f(z)|^2 t^{2k-1} dt.$$

When $k = 1$ we also write g in place of g_1. We remark that we have the pointwise inequality

$$(2.3.2) \qquad g_k(f,z) \leq C g_{k+1}(f,z).$$

To see this let $u(z,t) = T^t f(z)$; observe that all the t-derivatives of u tend to 0 as $t \to \infty$. Therefore, writing

$$(2.3.3) \qquad \partial_t^k u(z,t) = -\int_t^\infty \partial_s^{k+1} u(z,s) s^k s^{-k} ds,$$

and applying Holder's inequality we get

$$(2.3.4) \qquad |\partial_t^k u(z,t)|^2 \leq C t^{-2k+1} \left(\int_t^\infty |\partial_s^{k+1} u(z,s)|^2 s^{2k} ds \right).$$

This immediately gives the estimate

$$(2.3.5) \qquad (g_k(f,z))^2 \leq C \int_0^\infty \int_t^\infty |\partial_s^{k+1} u(z,s)|^2 s^{2k} ds dt.$$

An integration by parts now establishes the inequality (2.3.2). For the g_k functions we first prove an L^2 estimate.

Theorem 2.3.1. *For each $k \geq 1$ and $f \in L^2(\mathbb{C}^n)$*

$$\|g_k(f)\|_2^2 = 2^{-2k}\,\Gamma(2k)\|f\|_2^2.$$

Proof: From the definition it follows that

$$(2.3.6) \qquad \|g_k(f)\|_2^2 = \int_{\mathbb{C}^n} \int_0^\infty t^{2k-1} |\partial_t^k T^t f(z)|^2 dt\, dz.$$

We interchange the order of integration and use Plancherel formula (1.2.18) to calculate the inner integral. Since

$$(2.3.7) \qquad W(\partial_t^k T^t f) = W(f)(-1)^k H^k e^{-tH},$$

$$(2.3.8) \qquad \int_{\mathbb{C}^n} |\partial_t^k T^t f(z)|^2 dz = (2\pi)^{-n} \|W(f) H^k e^{-tH}\|_{HS}^2.$$

To calculate the right hand side we take the orthonormal basis $\{\Phi_\mu\}$ consisting of the Hermite functions. Observe that

$$\|W(f) H^k e^{-tH} \Phi_\mu\|_2^2 = (2|\mu|+n)^{2k} e^{-(2|\mu|+n)2t} \|W(f)\Phi_\mu\|_2^2.$$

This gives us

$$(2.3.9) \quad \|W(f) H^k e^{-tH}\|_{HS}^2 = \sum_\mu (2|\mu|+n)^{2k} e^{-2(2|\mu|+n)t} \|W(f)\Phi_\mu\|_2^2.$$

Integrating this against $t^{2k-1} dt$ we see that

$$(2.3.10) \qquad \|g_k(f,z)\|_2^2 = 2^{-2k}\,\Gamma(2k)(2\pi)^{-n} \sum_\mu \|W(f)\Phi_\mu\|_2^2$$

which equals $2^{-2k}\Gamma(2k)\|f\|_2^2$ again by Plancherel theorem. This proves the theorem. ∎

We next proceed to prove that the g-functions are bounded on $L^p(\mathbb{C}^n)$, $1 < p < \infty$. To do this we want to view the g function as an oscillatory singular integral with a kernel taking values in a Hilbert space. For ordinary singular integral operators taking values in a Hilbert space the following theorem is known (see Stein [S1]). In what follows E_1 and E_2 are two Hilbert spaces and $\mathcal{B}(E_1, E_2)$ is the Banach space of bounded operators from E_1 into E_2.

Theorem 2.3.2. Let $K(x)$ be a C^1 function away from the origin taking values in $\mathcal{B}(E_1, E_2)$. Assume that

(i) $\|K(x)\| \leq C|x|^{-n}$,

(ii) $\|\nabla K(x)\| \leq C|x|^{-n-1}$,

where $\|\cdot\|$ is the operator norm. Let f take values in H_1 and T be defined by
$$Tf(x) = p.v. \int_{\mathbb{R}^n} K(x-y)f(y)dy.$$
If T is bounded on $L^2(\mathbb{R}^n)$ then it is bounded also on $L^p(\mathbb{R}^n)$, $1 < p < \infty$.

We need an oscillatory version of the above theorem. To see how oscillatory singular integrals enter into the picture let us take a close look at the operator taking f into $\partial_t T^t f$. This is given by twisted convolution with a kernel k_t,

(2.3.11) $$k_t(z) = (2\pi)^{-n}(-1)\sum_{k=0}^{\infty}(2k+n)e^{-(2k+n)t}\varphi_k(z).$$

Taking E_2 to be Hilbert space $L^2(\mathbb{R}_+, tdt)$ we observe that

(2.3.12) $$(g(f,z))^2 = \|f \times k_t\|_{E_2}^2.$$

Thus, it is clear that we have to study oscillatory singular integrals taking values in the Hilbert space $L^2(\mathbb{R}_+, tdt)$. The following lemma shows that k_t is indeed a Calderon–Zygmund kernel.

Lemma 2.3.1. The kernel k_t satisfies

(i) $\|k_t(z)\|_{E_2} \leq C|z|^{-2n}$,

(ii) $\|\nabla k_t(z)\|_{E_2} \leq C|z|^{-2n-1}$.

Proof: Since the kernel k_t is defined by

(2.3.13) $$k_t(z) = (4\pi)^{-n}\frac{\partial}{\partial t}\left\{(\sinh t)^{-n}e^{-\frac{1}{4}|z|^2 \coth t}\right\}$$

it is easy to see that the following estimate holds:

(2.3.14) $$|k_t(z)| \leq Ct^{-n-1}(1+t^{-1}|z|^2)^{-n-1}.$$

From this it follows immediately that

$$(2.3.15) \quad \|k_t(z)\|_{E_2}^2 \leq C \int_0^\infty t^{-2n-1}(1+t^{-1}|z|^2)^{-2n-2} dt$$

$$\leq C|z|^{-2n} \int_0^\infty t^{-2n-1}(1+t^{-1})^{-2n-2} \leq C|z|^{-2n}$$

as the t integral is convergent. This proves (i). The proof of the second estimate is similar. Any derivative will in effect bring down a factor of $t^{-\frac{1}{2}}$ which accounts for the extra factor $|z|^{-1}$. The details are omitted. ∎

Let us take $E_1 = \mathbb{C}$ so that $k_t(z)$ can be thought of taking values in $\mathcal{B}(E_1, E_2)$. Define

$$(2.3.16) \quad Tf(z) = \int k_t(z-w) e^{-i/2 \operatorname{Im} z \cdot \overline{w}} f(w) dw$$

so that T is an oscillatory singular integral operator taking values in E_2. From Theorem 2.3.1 we know that T is bounded from $L^2(\mathbb{C}^n)$ into $L^2(\mathbb{C}^n, E_2)$. To show that T is bounded from $L^p(\mathbb{C}^n)$ into $L^p(\mathbb{C}^n, E_2)$ we decompose $Tf = T_0 f + T_\infty f$ as in the previous section. Following the arguments in the proof of Theorem 2.2.1 we can prove that T_0 is bounded from $L^p(\mathbb{C}^n)$ into $L^p(\mathbb{C}^n, E_2)$. The proof requires the result of Theorem 2.3.2 regarding the boundedness of vector valued singular integral operators.

To study T_∞ we take a partition of unity and write $T_\infty = \sum_{j=0}^\infty T_j$ as before. Again we want to look at the operator $T_j^* T_j$. The kernel of the operator T_j is given by $K_t^j(z-w) e^{-i/2 \operatorname{Im} z \cdot \overline{w}}$ where $K_t^j(z)$ stands for $k_t(z)\psi(2^{-j}z)$. Let $s_j(z,w)$ stand for the operator in $\mathcal{B}(E_1, E_2)$ which takes $\lambda \in E_1 = \mathbb{C}$ into $K_t^j(z-w) e^{-i/2 \operatorname{Im} z \cdot \overline{w}} \lambda$. Thus

$$(2.3.17) \quad T_j f(z) = \int s_j(z,w) f(w) dw.$$

Then it is easily verified that T_j^* is given by

$$(2.3.18) \quad T_j^* f(z) = \int s_j^*(w,z) f(w) dw$$

where $s_j^*(z,w)$ is the adjoint of the operator $s_j(z,w)$. Another easy calculation shows that for $g \in L^2(\mathbb{R}_+, tdt)$

$$(2.3.19) \quad s_j^*(z,w)g = \int_0^\infty K_t^j(z-w) e^{i/2 \operatorname{Im} z \cdot \overline{w}} g(t) t dt.$$

Finally, the kernel $L_j(z,w)$ of $T_j^* T_j$ is given by

$$(2.3.20) \quad L_j(z,w) = \iint_0^\infty K_t^j(\zeta - z) K_t^j(\zeta - w) e^{i/2 \operatorname{Im} \zeta \cdot (\bar{z}-\bar{w})} t \, dt \, d\zeta.$$

A change of variables in (2.3.20) shows that

$$(2.3.21) \quad L_j(z,w) = e^{-i/2 \operatorname{Im} z \cdot \bar{w}} G_j(z - w)$$

where the function G_j is scalar valued and defined by

$$(2.3.22) \quad G_j(z) = \iint_0^\infty \widetilde{K}_t^j(z - w) K_t^j(w) e^{i/2 \operatorname{Im} z \cdot \bar{w}} t \, dt \, dw$$

where we have written $\widetilde{K}_t^j(z) = K_t^j(-z)$. Thus one has the estimate

$$(2.3.23) \quad \|T_j^* T_j f\|_2 \leq C \|G_j\|_2 \|f\|_2.$$

It remains to get an estimate for $\|G_j\|_2$. Since G_j is given by

$$(2.3.24) \quad G_j(z) = \int_0^\infty \widetilde{K}_t^j \times K_t^j(z) t \, dt$$

applying Minkowski's integral inequality and using Theorem 1.2.2 we get

$$(2.3.25) \quad \|G_j\|_2 \leq C \int_0^\infty \|\widetilde{K}_t^j\|_2 \|K_t^j\|_2 t \, dt$$

which gives by Cauchy–Schwarz inequality

$$(2.3.26) \quad \|G_j\|_2 \leq C \Big(\int_0^\infty \|\widetilde{K}_t^j\|_2^2 t \, dt \Big)^{\frac{1}{2}} \Big(\int_0^\infty \|K_t^j\|_2^2 t \, dt \Big)^{\frac{1}{2}}.$$

Finally, making use of Lemma 2.3.1 and the support properties of K_t^j we get $\|G_j\|_2 \leq C 2^{-2nj}$. Thus we have shown that $\|T_j\|_2 \leq C 2^{-nj}$. Interpolating with the trivial estimate $\|T_j\|_1 \leq C$ we can conclude that $\|T_j\|_p \leq C 2^{-\epsilon j}$, $\epsilon > 0$. This proves that T_∞ is bounded from $L^p(\mathbb{C}^n)$ into $L^p(\mathbb{C}^n, E_2)$.

Theorem 2.3.3. *The g function satisfies*

$$C_1\|f\|_p \leq \|g(f)\|_p \leq C_2\|f\|_p.$$

Proof: We have proved that the oscillatory singular integral operator T defined in (2.3.16) is bounded from $L^p(\mathbb{C}^n)$ into $L^p(\mathbb{C}^n, E_2)$. Since

$$(g(f,z))^2 = \|Tf(z)\|_{E_2}^2$$

it follows that $\|g(f)\|_p \leq C_2\|f\|_p$. Now the reverse inequality can be deduced in a routine way using the result of Theorem 2.3.2. This completes the proof of Theorem 2.3.3. ∎

We need one more auxiliary function, namely the g_k^* function which is defined by

$$(2.3.27) \quad (g_k^*(f,z))^2$$
$$= \int_0^\infty \int_{\mathbb{C}^n} t^{1-n}(1+t^{-1}|z-w|^2)^{-k}|\partial_t T^t f(w)|^2 t\,dw\,dt.$$

Regarding this function we need the following result.

Theorem 2.3.4. *If $k > n$ and $p > 2$ one has*

$$\|g_k^*(f)\|_p \leq C\|f\|_p, \quad f \in L^p(\mathbb{C}^n).$$

Proof: As $k > n$ the function $t^{-n}(1+t^{-1}|z|^2)^{-k}$ is integrable. Therefore, if h is a nonnegative function it follows that

$$(2.3.28) \quad \int t^{-n}(1+t^{-1}|z-w|^2)^{-k} h(z)dz \leq CMh(w)$$

where Mh is the Hardy–Littlewood maximal function of h. Hence

$$(2.3.29) \quad \int (g_k^*(f,z))^2 h(z)dz \leq C \int (g(f,z))^2 Mh(z)dz$$

and by applying Holder's inequality and using the boundedness of g on M on L^p we get

$$(2.3.30) \quad \int g_k^*(f,z)^2 h(z)dz \leq C\|f\|_p^2 \|h\|_{q'}$$

where q' is the index conjugate to $q = \frac{p}{2}$. Taking supremum over all h with $\|h\|_{q'} \leq 1$ we complete the proof. ∎

2.4 Multipliers for Special Hermite Expansions

Our aim in this secction is to prove the following theorem regarding the multiplier transform T_m defined by

$$(2.4.1) \qquad T_m f = (2\pi)^{-\frac{n}{2}} \sum_\mu m(\mu) f \times \Phi_{\mu\mu}.$$

The conditions on m which ensure the boundedness of T_m on $L^p(\mathbb{C}^n)$ involve finite difference operators. We define Δ_j by

$$(2.4.2) \qquad \Delta_j m(\mu) = m(\mu + e_j) - m(\mu),$$

and define Δ_j^k inductively. If β is a multiindex we define Δ^β by

$$(2.4.3) \qquad \Delta^\beta m(\mu) = \Delta_1^{\beta_1} \Delta_2^{\beta_2} \cdots \Delta_n^{\beta_n} m(\mu).$$

The following is the multiplier theorem for the special Hermite expansions.

Theorem 2.4.1. *Assume that $k = n+1$ if n is odd and $k = n+2$ if n is even. Let m be a function defined on $\mathbb{N}^n$ which satisfies*

$$|\Delta^\beta m(\mu)| \leq C_\beta (1 + |\mu|)^{-|\beta|}$$

for all β with $|\beta| \leq k$. Then T_m is bounded on $L^p(\mathbb{C}^n)$, $1 < p < \infty$.

This theorem is proved using the results of the previous section in conjunction with the following proposition.

Proposition 2.4.1. *Assume that m and k are as in the previous theorem. Then the pointwise estimate $g_{k+1}(T_m f, z) \leq C g_k^*(f, z)$ holds.*

That the theorem follows from this proposition can be easily seen. In fact, if $p > 2$ using Theorems 2.3.3 and 2.3.4 one has

$$\|T_m f\|_p \leq C \|g_{k+1}(T_m f)\|_p \leq C \|g_k^*(f)\|_p \leq C \|f\|_p$$

which proves the theorem for $p > 2$. For $1 < p \leq 2$ we can apply a duality argument. Thus it remains to prove the above proposition. ∎

We define a kernel $M(z,t)$ by setting

$$(2.4.4) \qquad M(z,t) = (2\pi)^{-n} \sum_\mu m(\mu) e^{-(2|\mu|+n)t} \Phi_{\mu\mu}(z).$$

Let $u(z,t) = T^t f(z)$ and $U(z,t) = T^t(T_m f)$. Then it is easily verified that

(2.4.5) $$U(z, t+s) = u(z,s) \times M(z,t).$$

If we take k derivatives with respect to t and one derivative with respect to s and set $s = t$ we get the relation

(2.4.6) $$\partial_t^{k+1} U(z, 2t) = (\partial_t T^t f) \times \partial_t^k M(z).$$

The following lemma translates the conditions on the function m into properties of the kernel M.

Lemma 2.4.1. *Under the hypotheses of Theorem 2.4.1 the following estimates are true.*

(i) $|\partial_t^k M(z,t)| \leq C t^{-n-k}$,

(ii) $\displaystyle\int_{\mathbb{C}^n} |z|^{2k} |\partial_t^k M(z,t)|^2 dz \leq C t^{-n-k}$, *if k is even,*

(iii) $\displaystyle\int_{\mathbb{C}^n} |z|^{2k+2} |\partial_t^k M(z,t)|^2 dz \leq C t^{-n-k+1}$, *if k is odd.*

Postponing the proof of this lemma for a moment we will complete the proof of Proposition 2.4.1. In view of (2.4.6) one has

(2.4.7) $$|\partial_t^{k+1} T^{2t}(T_m f)(z)|^2$$
$$\leq \int_{\mathbb{C}^n} |\partial_t T^t f(w)| \, |\partial_t^k M(z-w, t)| dw$$
$$= A_t(z) + B_t(z)$$

where $A_t(z)$ and $B_t(z)$ are defined to be

(2.4.8) $$A_t(z) = \int_{|z-w|\leq t^{\frac{1}{2}}} |\partial_t T^t f(w)| \, |\partial_t^k M(z-w, t)| dw,$$

(2.4.9) $$B_t(z) = \int_{|z-w|> t^{\frac{1}{2}}} |\partial_t T^t f(w)| \, |\partial_t^k M(z-w, t)| dw.$$

Applying the Cauchy-Schwarz inequality and using (i) of the Lemma we get the estimate

(2.4.10) $$(A_t(z))^2 \leq \left(\int_{|z-w|\leq t^{\frac{1}{2}}} |\partial_t T^t f(w)|^2 dw \right)$$
$$\left(\int_{|z-w|\leq t^{\frac{1}{2}}} |\partial_t^k M(z-w, t)|^2 dw \right)$$
$$\leq C \int_{|z-w|\leq t^{\frac{1}{2}}} t^{-n-2k} |\partial_t T^t f(w)|^2 dw$$
$$\leq C t^{-n-2k} \int_{\mathbb{C}^n} (1+t^{-1}|z-w|^2)^{-k} |\partial_t T^t f(w)|^2 dw.$$

When k is even another application of Cauchy–Schwarz together with (ii) of the Lemma gives

$$(2.4.11) \quad (B_t(z))^2 \leq \left(\int_{|z-w|<t^{\frac{1}{2}}} |z-w|^{-2k} |\partial_t T^t f(w)|^2 dw \right)$$

$$\left(\int_{|z-w|<t^{\frac{1}{2}}} |z-w|^{2k} |\partial_t^k M(z-w,t)|^2 dw \right)$$

$$\leq C t^{-n-2k} \int_{\mathbb{C}^n} (1+t^{-1}|z-w|^2)^{-k} |\partial_t T^t f(w)|^2 dw.$$

When k is odd we use (iii) of the Lemma to get the same estimate for $B_t(z)$.

Thus we have obtained the estimate

$$(2.4.12) \quad |\partial_t^{k+1} T^{2t}(T_m f)(z)|^2$$

$$\leq C t^{-n-2k} \int_{\mathbb{C}^n} (1+t^{-1}|z-w|^2)^{-k} |\partial_t T^t f(w)|^2 dw.$$

Integrating this against $t^{2k-1} dt$ and recalling the definitions of g_{k+1} and g_k^* we immediately obtain the proposition. ∎

Let us now come to the proof of the lemma. Recall from (1.1.46) that the Laguerre polynomials L_k^{n-1} satisfy the generating function

$$(2.4.13) \quad \sum_{k=0}^{\infty} \frac{\Gamma(k+1)}{\Gamma(k+n)} (L_k^{n-1}(r))^2 e^{-r} w^k$$

$$= (1-w)^{-1}(-r^2 w)^{-(\frac{n-1}{2})} e^{-\frac{1}{2}\frac{1+w}{1-w}r} J_{n-1}\left(\frac{2(-r^2 w)^{\frac{1}{2}}}{1-w} \right).$$

Taking $w = e^{-2t}$ we have the formula

$$(2.4.14) \quad \sum_{k=0}^{\infty} \frac{\Gamma(k+1)}{\Gamma(k+n)} (L_k^{n-1}(r) e^{-\frac{1}{2}r})^2 e^{-(2k+n)t}$$

$$= \tfrac{1}{2}(\sinh t)^{-1} e^{-r \coth t} (ir)^{-(n-1)} J_{n-1}(ir \operatorname{cosech} t).$$

We also need the following estimates for the Bessel function J_{n-1}.

$$(2.4.15) \qquad |J_{n-1}(z)| \leq C|z|^\alpha, \quad |z| \leq 1,$$

$$(2.4.15') \qquad |J_{n-1}(iz)| \leq C z^{-\frac{1}{2}} e^z, \quad z \geq 1.$$

We can now prove part (ii) of the lemma. For the sake of simplicity of notation we assume that $m(\mu) = m(|\mu|)$. The general case is no different from this particular case.

Thus we need to estimate the function

(2.4.16) $\quad \partial_t^k M(z,t)$
$$= (2\pi)^{-n}(-1)^k \sum_{N=0}^{\infty}(2N+n)^k e^{-(2N+n)t} m(N)\varphi_N(z).$$

Applying the Cauchy–Schwarz inequality and noting that m is bounded we have

(2.4.17)
$$|\partial_t^k M(z,t)|^2 \leq C\Big(\sum_{N=0}^{\infty}(2N+n)^{2k}\frac{\Gamma(N+n)}{\Gamma(N+1)}e^{-(2N+n)t}\Big)$$
$$\cdot \Big(\sum_{N=0}^{\infty}\frac{\Gamma(N+1)}{\Gamma(N+n)}e^{-(2N+n)t}(\varphi_N(z))^2\Big).$$

As $\frac{\Gamma(N+n)}{\Gamma(N+1)} \leq CN^{n-1}$, the first factor on the right hand side of (2.4.17) is bounded by

(2.4.18) $\quad \displaystyle\sum_{N=0}^{\infty}(2N+n)^{2k+n-1}e^{-(2N+n)t}$

which equals $(-1)^{2k+n-1}$ times the $(2k+n-1)^{th}$ derivative of the function $e^{-nt}(1-e^{-2t})^{-1}$. And so (2.4.18) is bounded by a constant times t^{-2k-n}. Therefore, part (i) of the lemma will be proved once we show that the second factor on the right hand side of (2.4.17) is bounded by Ct^{-n}.

Writing $r = \frac{1}{2}|z|^2$, the formula (2.4.14) gives us

(2.4.19)
$$\sum_{N=0}^{\infty}\frac{\Gamma(N+1)}{\Gamma(N+n)}e^{-(2N+n)t}(\varphi_N(z))^2$$
$$= \tfrac{1}{2}(\sinh t)^{-1} e^{-r\coth t}(ir)^{-(n-1)}J_{n-1}(ir\operatorname{cosech} t).$$

When $r\operatorname{cosech} t \leq 1$ we use the estimate (2.4.15) to bound (2.4.19) by

(2.4.20) $\quad C(\sinh t)^{-1} r^{-(n-1)}(r\operatorname{cosech} t)^{n-1} \leq C(\sinh t)^{-n} \leq Ct^{-n}$.

When $r\operatorname{cosech} t > 1$ we use (2.4.15'). This gives the bound

(2.4.21) $\quad C(\sinh t)^{-1}e^{-r\coth t}r^{-(n-1)}r^{-\frac{1}{2}}(\sinh t)^{\frac{1}{2}}e^{r\operatorname{cosech} t}$
$$= C(\sinh t)^{-\frac{1}{2}}r^{-n+\frac{1}{2}}e^{-r\tanh t/2}.$$

Since $r > \sinh t$ the above is bounded by

(2.4.22) $\quad C(\sinh t)^{-n}e^{-r\tanh t/2} \le Ct^{-n}.$

This completes the proof of part (i) of the Lemma.

To prove parts (ii) and (iii) of the Lemma we need the following result. To state the result we need some more notation. Let Δ_+ and Δ_- denote the forward and backward finite difference operators defined by

(2.4.23) $\quad\begin{aligned}\Delta_+\psi(N) &= \psi(N+1) - \psi(N),\\ \Delta_-\psi(N) &= \psi(N) - \psi(N-1).\end{aligned}$

Consider a kernel of the form

$$(2.4.24) \quad M_\psi(z) = \sum_{N=0}^{\infty} \psi(N)\varphi_N(z).$$

The following lemma gives the effect of multiplying $M_\psi(z)$ by $\frac{1}{2}|z|^2$.

Lemma 2.4.2.

$$\tfrac{1}{2}|z|^2 M_\psi(z) = -\sum_{N=0}^{\infty}(N\Delta_-\Delta_+\psi(N) + n\Delta_-\psi(N))\varphi_N(z).$$

Proof: We make use of the recursion formula (1.1.37) which gives us the relation

(2.4.25) $\quad rL_N^{n-1}(r) = (2N+n)L_N^{n-1}(r) - (N+1)L_{N+1}^{n-1}(r)$
$$- (N+n-1)L_{N-1}^{n-1}(r).$$

Using this and rearranging we get the Lemma. ∎

Coming back to the proof of Lemma 2.4.1 we let

(2.4.26) $\quad \psi(N) = (2N+n)^k e^{-(2N+n)t}m(N)$

which satisfies an estimate of the form

(2.4.27) $\quad |\psi(N)| \le C(2N+n)^k e^{-(2N+n)t}.$

We also have $\partial_t^k M(z,t) = \sum_{N=0}^{\infty} \psi(N)\varphi_N(z)$. In view of the preceding lemma one has

(2.4.28) $$\tfrac{1}{2}|z|^2 \partial_t^k M(z,t) = \sum_{N=0}^{\infty} \psi_1(N)\varphi_N(z)$$

where the function ψ_1 is related to ψ by

(2.4.29) $$\psi_1(N) = -(N\Delta_-\Delta_+\psi(N) + n\Delta_-\psi(N)).$$

Using the Leibnitz formula for the finite difference operators, and the hypothesis on the function m we can conclude that there is an $\varepsilon_1 > 0$ such that

(2.4.30) $$|\psi_1(N)| \leq C(2N+n)^{k-1} e^{-\varepsilon_1(2N+n)t}.$$

Assuming k is even, iteration of the above proves that

(2.4.31) $$2^{-k/2}|z|^k \partial_t^k N(z,t) = \sum_{N=0}^{\infty} \psi_{k/2}(N)\varphi_N(z)$$

where the function $\psi_{k/2}(N)$ satisfies the estimate

(2.4.32) $$|\psi_{k/2}(N)| \leq C(2N+n)^{k/2} e^{-\varepsilon(2N+n)t}$$

where $\varepsilon > 0$ is some constant.

Finally, the orthogonality properties of the function φ_N prove that

(2.4.33) $$2^{-k}\int |z|^{2k}|\partial_t^k M(z,t)|^2 dz = \sum_{N=0}^{\infty} |\psi_{k/2}(N)|^2 \int_{\mathbb{C}^n} (\varphi_N(z))^2 dz.$$

As $\|\varphi_N\|_2 \leq CN^{\frac{n-1}{2}}$ the above gives the estimate

(2.4.34) $$\int |z|^{2k}|\partial_t^k N(z,t)|^2 dz \leq C \sum_{N=0}^{\infty} (2N+n)^{k+n-1} e^{-\varepsilon(2N+n)t}$$

where we have made use of the estimate (2.4.32). The last sum is clearly bounded by Ct^{-n-k}. This completes the proof of part (ii) of Lemma 2.4.1. Part (iii) is proved in a similar way. ∎

2.5 Riesz Means for Special Hermite Expansions

In this section we study the almost everywhere and mean convergence of the Riesz means defined in (2.1.8). Whenever we want to estimate the kernel, it is convenient to consider the Cesàro means (2.1.7). There is a formula due to Gergen [Ge] connecting the kernel of the two means. If σ_N^δ is the kernel of the Cesàro means and S_R^δ is that of the Riesz means then

$$(2.5.1) \qquad S_R^\delta(z) = R^{-\delta} \sum_{k=0}^{R} v(R-k) A_k^\delta \sigma_k^\delta(z)$$

where the function v satisfies $|v(t)| \leq C(1+t^2)^{-1}$. Regarding the kernel σ_N^δ we first prove the following proposition.

Proposition 2.5.1. *The kernel σ_N^δ satisfies*

$$|\sigma_N^\delta(z)| \leq CN^n(1 + N^{\frac{1}{2}}|z|)^{-\delta-n-\frac{1}{3}}.$$

When $\delta > n - \frac{1}{2}$ the kernel σ_N^δ is uniformly integrable in the sense that

$$\int_{\mathbb{C}^n} |\sigma_N^\delta(z)| dz \leq C$$

where C is a constant independent of N.

Proof: The kernel of the Cesàro means is given by

$$(2.5.2) \qquad \sigma_N^\delta(z) = \frac{(2\pi)^{-n}}{A_N^\delta} \sum_{k=0}^{N} A_{N-k}^\delta L_k^{n-1}(\tfrac{1}{2}|z|^2) e^{-\frac{1}{4}|z|^2}.$$

If we use the formula (1.1.38) it is immediate that

$$(2.5.3) \qquad \sigma_N^\delta(z) = \frac{(2\pi)^{-n}}{A_N^\delta} L_N^{\delta+n}(\tfrac{1}{2}|z|^2) e^{-\frac{1}{4}|z|^2}.$$

We can now appeal to the asymptotic estimates stated in Lemma 1.5.3. Using those estimates it is not difficult to prove that the above estimate for the kernel is valid. Similarly, when $\delta > n - \frac{1}{2}$ it can be checked that σ_N^δ is uniformly integrable. The proofs are straightforward and so we leave the details. ∎

Regarding the almost everywhere and mean convergence of S_R^δ we have the following theorem.

Theorem 2.5.1.
(i) If $\delta > n - \frac{1}{2}$ the Riesz means S_R^δ are uniformly bounded on $L^p(\mathbb{C}^n)$, $1 \leq p \leq \infty$. If $1 \leq p < \infty$, $S_R^\delta f$ converges to f in the norm.
(ii) If $\delta > n - \frac{1}{3}$ then $S_R^\delta f(z)$ converges to $f(z)$ almost everywhere.

Proof: It is enough to prove the theorem for the Cesàro means. When $\delta > n - \frac{1}{2}$ it follows from the preceding proposition that σ_N^δ are uniformly bounded on $L^p(\mathbb{C}^n)$, $1 \leq p \leq \infty$. A density argument proves that $\sigma_N^\delta f$ converges to f in the norm. When $\delta > n - \frac{1}{3}$ it follows that

$$(2.5.4) \qquad |\sigma_N^\delta f(z)| \leq CN^n \int (1 + N^{\frac{1}{2}}|w|)^{-\delta - n - \frac{1}{3}} |f(z - w)| dw$$
$$\leq CMf(z)$$

where M is the Hardy–Littlewood maximal function. This shows that the maximal operator $\sigma_*^\delta f(z) = \sup_{N>0} |\sigma_N^\delta f(z)|$ is of weak type $(1,1)$. The almost everywhere convergence follows form this. The proof is standard and so can be omitted. ∎

The almost everywhere convergence result in the above theorem can be improved. In fact, using only the asymptotic estimates of the Laguerre functions we are able to prove the following result.

Theorem 2.5.2. Assume that $\delta > n - \frac{1}{2}$, $p > \frac{4}{3}$ and $f \in L^p(\mathbb{C}^n)$. Then $S_R^\delta f$ converges to f almost everywhere.

Proof: Again we consider only the Cesàro means. The kernel $\sigma_N^\delta(z)$ satisfies

$$(2.5.5) \qquad |\sigma_N^\delta(z)| \leq CN^{\frac{n-\delta}{2}} r^{-\frac{(\delta+n)}{2}} \mathcal{L}_N^{\delta+n}(r)$$

where $\mathcal{L}_k^\alpha$ are the normalized Laguerre functions and $r = \frac{1}{2}|z|^2$. We write $\sigma_N^\delta(z) = \sum_{j=1}^3 G_N^j(z)$ where G_N^j are defined as follows. Let $\nu = 4N + 2(\delta + n) + 2$.

$$(2.5.6) \qquad G_N^1(z) = \sigma_N^\delta(z) \mathcal{X}\left(0 \leq r \leq \frac{\nu}{2}\right),$$

$$(2.5.6') \qquad G_N^2(z) = \sigma_N^\delta(z) \mathcal{X}\left(\frac{\nu}{2} < r \leq \frac{3\nu}{2}\right).$$

Using the asymptotic estimates for the Laguerre functions it is not difficult to show that

$$(2.5.7) \qquad |G_N^1(z)| \leq CN^n (1 + N^{\frac{1}{2}}|z|)^{-\delta - n - \frac{1}{2}}.$$

This proves $\sup_{N>0} |G_N^1 \times f(z)|$ is of weak type $(1,1)$ provided $\delta > n - \frac{1}{2}$. Similarly we can show that $\sup_{N>0} |G_N^3 \times f(z)|$ is also of weak type $(1,1)$.

We will show that $\sup_{N>0} |G_N^2 \times f(z)|$ is of weak type (p,p) for any $p > \frac{4}{3}$. This is a bit tricky and we have to use the condition $p > \frac{4}{3}$. In the interval $\frac{\nu}{2} < r \leq \frac{3\nu}{2}$ we have the estimate

$$(2.5.8) \qquad |\mathcal{L}_N^{n+\delta}(r)| \leq C\nu^{-\frac{1}{4}}(1 + |r - \nu|)^{-\frac{1}{4}}.$$

Now one has

$$(2.5.9) \qquad |G_N^2 \times f(z)| \leq \int |G_N^2(w)| \, |f(z - w)| dw$$

which gives by Holder's inequality

$$(2.5.10) \qquad |G_N^2 \times f(z)|$$
$$\leq \Big(\int_{|w|^2 \leq \frac{3}{2}\nu} |f(z-w)|^p dw\Big)^{\frac{1}{p}} \Big(\int |G_N^2(w)|^{p'} dw\Big)^{\frac{1}{p'}}.$$

The first factor on the right hand side is bounded by a constant times

$$(2.5.11) \qquad \nu^{n/p} \Big(\frac{1}{\nu^n} \int_{|w|^2 \leq \frac{3}{2}\nu} |f(z-w)|^p dw\Big)^{\frac{1}{p}} \leq C\nu^{\frac{n}{p}} (M|f|^p)^{\frac{1}{p}}.$$

On the other hand in view of (2.5.8) and (2.5.5) the second factor is bounded by

$$(2.5.12) \qquad C\nu^{-\delta-\frac{1}{4}} \Big(\int_{\nu/2}^{3\nu/2} (1 + |r - \nu|)^{-p'/4} r^{n-1} dr\Big)^{\frac{1}{p'}}$$
$$\leq C\nu^{-\delta - \frac{1}{4} + \frac{n-1}{p'}} \Big(\int_0^{\nu/2} (1 + r)^{-p'/4} dr\Big)^{\frac{1}{p'}}.$$

As $p > 4/3$ the above integral converges and so we have the bound $C\nu^{-\delta - \frac{1}{2} + \frac{n}{p'}}$. Hence

$$(2.5.13) \qquad |G_N^2 \times f(z)| \leq C\nu^{-\delta - \frac{1}{2} + n}(M|f|^p)^{\frac{1}{p}}.$$

As $\delta > n - \frac{1}{2}$, $\sup_{N>0} |G_N^2 \times f(z)| \leq C(M|f|^p)^{\frac{1}{p}}$ and this proves that σ_*^δ is weak type (p,p) for all $p > \frac{4}{3}$. Hence the theorem follows. ∎

The rest of this section is devoted to the study of the boundedness properties of the Riesz means S_R^δ when $0 < \delta \leq n - \frac{1}{2}$. Using a transplantation theorem it can be proved that S_R^δ cannot be uniformly bounded on L^p unless p lies in the interval $\frac{4n}{2n+1+2\delta} < p < \frac{4n}{2n-1-2\delta}$. Again we remark that this can be proved without recourse to the transplantation theorem. In fact, when f is a radial function it can be shown that the special Hermite expansion reduces to a Laguerre expansion (see Chapter 6). Then using certain lower bounds for the L^p norms of the Laguerre functions we can prove that the above condition on p is necessary for S_R^δ to be uniformly bounded on $L^p(\mathbb{C}^n)$ (see the arguments given in the case of Hermite series in Chapter 5).

We therefore make the conjecture that when $0 < \delta \leq n - \frac{1}{2}$, the Riesz means S_R^δ are uniformly bounded on $L^p(\mathbb{C}^n)$ whenever p lies in the critical range. In the case of radial functions the conjecture will be proved in Chapter 6. In the general case the conjecture is proved only for $\delta > \frac{1}{2}$. Let us define, for $1 \leq p \leq 2$, $\delta(p) = \max\{2n(\frac{1}{p} - \frac{1}{2}) - \frac{1}{2}, 0\}$ to be the critical index for the L^p summability of the special Hermite expansion. We have the following result.

Theorem 2.5.3. *If $1 \leq p \leq \frac{2n}{n+1}$ and $\delta > \delta(p)$, then S_R^δ are uniformly bounded on $L^p(\mathbb{C}^n)$.*

Corollary 2.5.1. *Assume that $\delta > \frac{1}{2}$. Then S_R^δ are uniformly bounded on $L^p(\mathbb{C}^n)$, $\frac{4n}{2n+1+2\delta} < p < \frac{4n}{2n-1-2\delta}$.*

Before going into the details we want to make the following remarks concerning the proof of Theorem 2.5.3. We prove the theorem using the socalled method of "restriction theorems". The basic idea goes back to the work of Fefferman–Stein [Fe2] where they studied the Riesz means for the Fourier transform. There they showed that in order to prove the uniform boundedness of the Riesz means it is enough to have certain $L^p - L^2$ restriction theorems for the spectral projections. This method has been further developed by Sogge [So] where he has applied it to study the Riesz means for the eigenfunction expansions associated to second order elliptic differential operators on a compact manifold.

The two ingredients that are necessary to carry out the above method of proof are (i) an estimate for the kernel $s_R^\delta(z)$ of the Riesz means, which is provided by Proposition 2.5.1 and (ii) the $L^p - L^2$ restriction theorem stated below. Let $Q_k f = (2\pi)^{-n} f \times \varphi_k$ be the projection associated to the kth eigenspace of the operator L.

Theorem 2.5.4. *For $1 \leq p \leq \frac{2n}{n+1}$ the projections Q_k satisfy the estimates*
$$\|Q_k f\|_2 \leq C k^{n(\frac{1}{p} - \frac{1}{2}) - \frac{1}{2}} \|f\|_p, \quad f \varepsilon L^p(\mathbb{C}^n).$$

The above theorem will be proved in the next section. We assume the theorem and proceed to the proof of Theorem 2.5.3. We take a partition of unity and decompose S_R^δ into several pieces. Let $\varphi \in C_0^\infty(\frac{1}{2}, 2)$ be a non-negative function with the property that $\sum_{j=-\infty}^{\infty} \varphi(2^j t) = 1$ for $t > 0$. For each j we set

(2.5.14) $$\varphi_{R,j}^\delta(t) = \left(1 - \frac{t}{R}\right)^\delta \varphi\left(2^j \left(1 - \frac{t}{R}\right)\right)$$

and define for $j = 1, 2, \ldots$

(2.5.15) $$S_{R,j}^\delta f = \sum_{k=0}^{\infty} \varphi_{R,j}^\delta(2k+n) Q_k f.$$

When $j = 0$ we define

(2.5.16) $$S_{R,0}^\delta f = \sum_{k=0}^{\infty} \left(1 - \frac{2k+n}{R}\right)^\delta \varphi_0\left(1 - \frac{2k+n}{R}\right) Q_k f$$

where $\varphi_0(t) = 1 - \sum_{j=1}^{\infty} \varphi(2^j t)$. Then we have

(2.5.17) $$S_R^\delta f = \sum_{j=0}^{[\log \sqrt{R}]} S_{R,j}^\delta f + T_R^\delta f.$$

It is easy to see that the function m defined by

$$m(k) = \left(1 - \frac{2k+n}{R}\right)^\delta \varphi_0\left(1 - \frac{2k+n}{R}\right)$$

satisfies the conditions of the multiplier Theorem 2.4.1 uniformly in R. It follows that $S_{R,0}^\delta$ is uniformly bounded on $L^p(\mathbb{C}^n)$, $1 < p < \infty$. For each $j = 1, 2, \ldots$ we show that there exists an $\varepsilon > 0$ such that

(2.5.18) $$\|S_{R,j}^\delta f\|_p \leq C 2^{-\varepsilon j} \|f\|_p.$$

We also prove the same estimate for the remainder term T_R^δ. Theorem 2.5.3 will be proved by summing a geometric series.

To establish (2.5.18) we split the kernel $s_R^\delta(z)$ into two parts. Since $\delta > \delta(p)$ it follows that for some $\gamma > 0$ we also have

$$\delta + \tfrac{1}{2} > (1+\gamma)\bigl(\delta(p) + \tfrac{1}{2}\bigr).$$

Fix a γ with this property and write $s_R^\delta(z) = k_1^j(z) + k_2^j(z)$ where

(2.5.19) $$k_1^j(z) = s_{R,j}^\delta(z) \mathcal{X}(|z| \leq 2^{j(1+\gamma)} R^{-\frac{1}{2}}).$$

Let K_1^j and K_2^j be the operators corresponding to the kernels k_1^j and k_2^j. The operator K_2^j is taken care of in the following proposition.

Proposition 2.5.2. *Given $\gamma > 0$ there is an $\varepsilon > 0$ such that*

$$\|K_2^j f\|_p \leq C 2^{-\varepsilon j} \|f\|_p.$$

Proof: It is for the proof of this proposition we need a pointwise estimate for the kernel $s_R^\delta(z)$. In view of Proposition 2.5.1 and the formula (2.5.1) it follows that s_R^δ satisfies the estimate

(2.5.20) $\qquad |s_R^\delta(z)| \leq CR^n (1 + R^{\frac{1}{2}}|z|)^{-\delta - n - \frac{1}{3}}.$

We need this estimate only when δ is large. We will show that there is an $\varepsilon > 0$ such that

(2.5.21) $\qquad \int_{\mathbb{C}^n} |k_2^j(z)| dz \leq C \, 2^{-\varepsilon j}.$

This will immediately prove the proposition.

To prove (2.5.21) we need to recall a few elementary facts concerning Riesz means. First of all if we let $S(t, z) = S_t^0(z)$ is the kernel of the partial sum operator, then $t \to S(t, z)$ is a function of bounded variation and for $j = 1, 2, \ldots, [\log \sqrt{R}]$ one has

(2.5.22) $\qquad s_{R,j}^\delta(z) = -\int_{-\infty}^{\infty} \varphi_{R,j}^\delta(t) dS(t, z).$

We want to integrate by parts using this identity. We observe that for each $m \geq 1$

(2.5.23) $\qquad \dfrac{\partial}{\partial t}(t^m S_t^m(z)) = m t^{m-1} S_t^{m-1}(z).$

Consequently, the last two identities imply that

(2.5.24) $\qquad s_{R,j}^\delta(z) = c_n \int_{-\infty}^{\infty} t^{m-1} s_t^{m-1}(z) \partial_t^m \varphi_{R,j}^\delta(t) dt.$

It is now easy to finish the proof. In fact since $\varphi_{R,j}^\delta$ is supported in $(\frac{1}{2}R, R)$ we have

(2.5.25) $\qquad |\partial_t^m \varphi_{R,j}^\delta(t)| \leq C_m (2^j R^{-1})^m.$

If we use this together with the estimate $|s_R^{m-1}(z)| \leq CR^n(1 + R^{\frac{1}{2}}|z|)^{-m}$ which is weaker than (2.5.20) we immediately get

(2.5.26) $\qquad |s_{R,j}^\delta(z)| \leq CR^n 2^{jn} (R^{\frac{1}{2}}|z|)^{-m}.$

This estimate shows that

$$(2.5.27) \qquad \int |k_2^j(z)|dz \leq C 2^{jn} \int_{2^j(1+\gamma)}^{\infty} r^{-m+n-1} dr.$$

Choosing m large we can ensure that (2.5.21) is valid with some $\varepsilon > 0$. ∎

To study the operator K_1^j we need the following proposition which makes use of the restriction theorem.

Proposition 2.5.3. *There is an $\varepsilon > 0$ such that*

$$\|S_{R,j}^\delta f\|_{L^p(B)} \leq C 2^{-\varepsilon j} \|f\|_p$$

for any ball B of radius $2^{j(1+\gamma)} R^{-\frac{1}{2}}$.

Proof: Applying Holder's inequality one has

$$(2.5.28) \qquad \|S_{R,j}^\delta f\|_{L^p(B)} \leq |B|^{\frac{1}{p}-\frac{1}{2}} \|S_{R,j}^\delta f\|_2$$

where $|B|$ is the volume of the ball B. Since

$$(2.5.29) \qquad S_{R,j}^\delta f = \sum \varphi_{R,j}^\delta(2k+n) Q_k f,$$

using the orthogonality of the projections Q_k it follows that

$$(2.5.30) \qquad \|S_{R,j}^\delta f\|_2^2 = \sum |\varphi_{R,j}^\delta(2k+n)|^2 \|Q_k f\|_2^2.$$

Now, as the function $\varphi_{R,j}^\delta$ is supported in the interval $R(1-2^{-j+1}) \leq t \leq R(1-2^{-j-1})$ we have the estimate $|\varphi_{R,j}^\delta(t)| \leq C 2^{-j\delta}$.

If we use the above estimate for $\varphi_{R,j}^\delta$ and the restriction theorem for $Q_k f$ it follows that

$$(2.5.31) \qquad \|S_{R,j}^\delta f\|_2^2 \leq C 2^{-2j\delta} \sum k^{\delta(p)-\frac{1}{2}},$$

the sum being extended over $R(1-2^{-j+1}) \leq (2k+n) \leq R(1-2^{-j-1})$. Thus we have

$$(2.5.32) \qquad \|S_{R,j}^\delta f\|_2^2 \leq C 2^{-2j(\delta+\frac{1}{2})} R^{\delta(p)+\frac{1}{2}}.$$

58   CHAPTER 2

Using this and the estimate $|B| \leq C 2^{2n(1+\gamma)j} R^n$ in (2.5.28) we finally obtain

(2.5.33) $\qquad \|S^\delta_{R,j} f\|_{L^p(B)} \leq C\, 2^{-j(\delta+\frac{1}{2})} 2^{j(1+\gamma)(\delta(p)+\frac{1}{2})} \|f\|_p.$

This proves the proposition by the choice of γ. ∎

We are now in a position to complete the proof of Theroem 2.5.3. To deal with K_1^j we decompose f into three parts. Let $\zeta \in \mathbb{C}^n$ and define

(2.5.34) $\qquad f_1(z) = f(z) \mathcal{X}\left(|z-\zeta| \leq \tfrac{3}{4} 2^{j(1+\gamma)} R^{-\frac{1}{2}}\right),$

(2.5.34') $\qquad f_2(z) = f(z) \mathcal{X}\left(\tfrac{3}{4} 2^{j(1+\gamma)} R^{-\frac{1}{2}} < |z-\zeta| \leq \tfrac{5}{4} 2^{j(1+\gamma)} R^{-\frac{1}{2}}\right)$

and $f_3 = f - f_1 - f_2$. Let $B(\zeta)$ be the ball $|z-\zeta| \leq \tfrac{1}{4} 2^{j(1+\gamma)} R^{-\frac{1}{2}}$. We will show that

(2.5.35) $\qquad \int_{B(\zeta)} |K_1^j f(z)|^p dz \leq C\, 2^{-\epsilon jp} \int_{|z-\zeta| \leq \frac{5}{4} 2^{j(1+\gamma)} R^{-\frac{1}{2}}} |f(z)|^p dz.$

Integration with respect to ζ will prove

(2.5.36) $\qquad \int |K_1^j(f(z))|^p dz \leq C 2^{-\epsilon jp} \|f\|_p^p.$

When $|z-\zeta| \leq \tfrac{1}{4} 2^{j(1+\gamma)} R^{-\frac{1}{2}}$ and w belonging to the support of f_3 it follows that $|z-w| > 2^{j(1+\gamma)} R^{-\frac{1}{2}}$ and consequently $K_1^j f_3 = 0$. When $|z-\zeta| \leq \tfrac{1}{4} 2^{j(1+\gamma)} R^{-\frac{1}{2}}$ and w belonging to the support of f_2 one has $|z-w| > \tfrac{1}{2} 2^{j(1+\gamma)} R^{-\frac{1}{2}}$ and we can repeat the proof of Proposition 2.5.2 to conclude that

(2.5.37) $\qquad \int_{B(\zeta)} |K_1^j f_2(z)|^p dz \leq C 2^{-\epsilon jp} \int |f_2(z)|^p dz.$

Finally, applying Proposition 2.5.3 to $K_1^j f_1$ we obtain the estimate

(2.5.38) $\qquad \int_{B(\zeta)} |K_1^j f_1(z)|^p dz \leq C 2^{-\epsilon jp} \int |f_1(z)|^p dz.$

The last two estimates together prove (2.5.36). This completes the proof of Theorem 2.5.3 and the corollary is immediate. ∎

2.6 Proof of the Restriction Theorem

In this section we prove Theorem 2.5.4. We embed the projections Q_k into an analytic family of operators G_k^α defined for $\operatorname{Re} \alpha > -\frac{1}{2n}$ and then apply analytic interpolation theorem. To find the right G_k^α we make use of the formula (1.1.41) connecting the Laguerre polynomial L_k^α with the Hermite polynomial H_{2k}. If we let

$$(2.6.1) \qquad \widetilde{\psi}_k^\alpha(x) = \frac{\Gamma(k+1)\Gamma(\alpha+1)}{\Gamma(k+\alpha+1)} L_k^\alpha(x^2) e^{-\frac{1}{2}x^2}, \quad x \geq 0$$

it follows from (1.1.41) that one has the formula

$$(2.6.2) \qquad \widetilde{\psi}_k^\alpha(x) = \frac{(-1)^k \sqrt{\pi}\, \Gamma(k+1)\Gamma(\alpha+1)}{\Gamma(\alpha+\frac{1}{2})\Gamma(2k+1)}$$
$$\times \int_{-1}^{1} (1-t^2)^{\alpha-\frac{1}{2}} H_{2k}(xt) e^{-\frac{1}{2}x^2}\, dt.$$

The above formula is valid even if α is complex provided $\operatorname{Re} \alpha > -\frac{1}{2}$. Using the same notation for the function $\widetilde{\psi}_k^\alpha(\frac{1}{2}|z|^2)$ we consider the family of operators

$$(2.6.3) \qquad G_k^\alpha f(z) = f \times \widetilde{\psi}_k^{n\alpha}(z).$$

Then G_k^α is an analytic family of operators and one has the relation

$$(2.6.4) \qquad G_k^{\frac{n-1}{n}} f(z) = \frac{\Gamma(n)\Gamma(k+1)}{\Gamma(k+n)} f \times \varphi_k(z).$$

We will show that G_k^α is an admissible family of operators.

Proposition 2.6.1. Let $\alpha = \sigma + i\tau$. The function $\widetilde{\psi}_k^\alpha$ satisfies the estimate
$$|\widetilde{\psi}_k^\alpha(x)| \leq C(1+|\tau|)^{\frac{1}{2}}$$
for all $x \geq 0$ uniformly in $0 \leq \sigma \leq n$.

Proof: From (2.6.2) it follows that

$$(2.6.5) \qquad |\widetilde{\psi}_k^\alpha(x)| \leq C \left| \frac{2^k \Gamma(k+1)\Gamma(\alpha+1)}{\Gamma(\alpha+\frac{1}{2})(\Gamma(2k+1))^{\frac{1}{2}}} \right| \left| \int_{-1}^{1} (1-t^2)^{\frac{1}{2}} |h_{2k}(xt)|\, dt \right|$$

where h_{2k} is the $2k$th Hermite function. It is therefore enough to prove the estimates

$$\left|\frac{2^k\Gamma(\alpha+1)\Gamma(k+1)}{\Gamma(\alpha+\frac{1}{2})(\Gamma(2k+1))^{\frac{1}{2}}}\right| \leq C(2k+1)^{\frac{1}{4}}(1+|\tau|)^{\frac{1}{2}}, \tag{2.6.6}$$

$$\int_{-1}^{1}(1-t^2)^{-\frac{1}{2}}|h_{2k}(xt)|dt \leq C(2k+1)^{-\frac{1}{4}}. \tag{2.6.6'}$$

Using Stirling's formula for the Γ function we observe that

$$2^k\Gamma(k+1) \leq C(2k+1)^{\frac{1}{4}}(\Gamma(2k+1))^{\frac{1}{2}}. \tag{2.6.7}$$

To estimate $\Gamma(\alpha+1)/\Gamma(\alpha+\frac{1}{2})$ we use Legendre's duplication formula, viz.

$$\sqrt{\pi}\Gamma(2\alpha) = 2^{2\alpha-1}\Gamma(\alpha)\Gamma(\alpha+\tfrac{1}{2}). \tag{2.6.8}$$

In view of this formula one has

$$\sqrt{\pi}\Gamma(\alpha+1)/\Gamma\left(\alpha+\tfrac{1}{2}\right) = 2^{2\alpha-1}\alpha(\Gamma(\alpha))^2/\Gamma(2\alpha). \tag{2.6.9}$$

Using Stirling's formula again we conclude that

$$|\Gamma(\alpha+1)| \leq C(1+|\tau|)^{\frac{1}{2}}|\Gamma(\alpha+\tfrac{1}{2})|. \tag{2.6.10}$$

This proves (2.6.6). The estimate (2.6.6') is proved in the following Lemma. ∎

Lemma 2.6.1.

$$\int_{-1}^{1}(1-t^2)^{-\frac{1}{2}}|h_n(xt)|dt \leq C(2n+1)^{-\frac{1}{4}}.$$

Proof: Let us write $N = (2n+1)$. By making a change of variables it is enough to show that for any $x > 0$

$$\int_{0}^{x}(x^2-t^2)^{-\frac{1}{2}}|h_n(t)|dt \leq CN^{-\frac{1}{4}}. \tag{2.6.11}$$

To do this we need the asymptotic properties of the Hermite functions stated in Lemma 1.5.1. We need to establish

$$I = \int_{a}^{x}(x^2-t^2)^{-\frac{1}{2}}|h_n(t)|dt \leq CN^{-\frac{1}{4}} \tag{2.6.12}$$

where $a < x \leq b$ for various values of a and b. We have to deal with several cases.

Case (i). $a = 0, b = \frac{1}{2}N^{\frac{1}{2}}$.
In this case we have the estimate $|h_n(t)| \leq CN^{-\frac{1}{6}}(N^{\frac{1}{2}}-t)^{-\frac{1}{4}} \leq CN^{-\frac{1}{4}}$.
Therefore, it follows that

$$(2.6.13) \qquad I \leq CN^{-\frac{1}{4}} \int_0^x (x^2 - t^2)^{-\frac{1}{2}} dt \leq CN^{-\frac{1}{4}}.$$

Case (ii). $a = \frac{1}{2}N^{\frac{1}{2}}, b = N^{\frac{1}{2}} - N^{-\frac{1}{6}}$.
In this case we use the same estimate as in the previous case. Thus we have

$$(2.6.14) \qquad I \leq CN^{-\frac{1}{6}} \int_a^x (x^2 - t^2)^{-\frac{1}{2}} (N^{\frac{1}{2}} - t)^{-\frac{1}{4}} dt.$$

Since $\frac{1}{2}N^{\frac{1}{2}} \leq x < N^{\frac{1}{2}}$ and $N^{\frac{1}{2}} - t \geq x - t$ we get

$$I \leq CN^{-\frac{3}{8}} \int_0^x (x-t)^{-\frac{3}{4}} dt \leq CN^{-\frac{3}{8}} x^{\frac{1}{4}} \leq CN^{-\frac{1}{4}}.$$

Case (iii). $a = N^{\frac{1}{2}} - N^{-\frac{1}{6}}, b = N^{\frac{1}{2}} + N^{-\frac{1}{6}}$.
In this case we have $|h_n(t)| \leq CN^{-\frac{1}{12}}$ and so we immediately get the estimate

$$(2.6.15) \qquad I \leq CN^{-\frac{1}{12}} N^{-\frac{1}{4}} (x-a)^{\frac{1}{2}}.$$

Since $x - a = x - N^{\frac{1}{2}} + N^{-\frac{1}{6}} \leq 2N^{-\frac{1}{6}}$ we get $I \leq CN^{-\frac{1}{4}}$.

Case (iv). $a = N^{\frac{1}{2}} + N^{-\frac{1}{6}}, b = (2N)^{\frac{1}{2}}$.
In this case there is a positive number ε such that

$$(2.6.16) \qquad |h_n(t)| \leq CN^{-\frac{1}{8}}(t - N^{\frac{1}{2}})^{-\frac{1}{4}} \exp\left(-\varepsilon N^{\frac{1}{4}}(t - N^{\frac{1}{2}})^{\frac{3}{2}}\right).$$

Therefore, the integral I is bounded by

$$(2.6.17) \qquad I \leq CN^{-\frac{1}{8}} \int_a^x (x^2 - t^2)^{-\frac{1}{2}} (t - N^{\frac{1}{2}})^{-\frac{1}{4}} e^{-\varepsilon N^{\frac{1}{4}}(t-N^{\frac{1}{2}})^{\frac{3}{2}}} dt.$$

Applying Holder's inequality with $\frac{4}{3} < p < 2$ we see that $I \leq CN^{-\frac{1}{8}} AB$ where

$$(2.6.18) \qquad A = \left(\int_a^x (x^2 - t^2)^{-p/2} dt \right)^{\frac{1}{p}} \leq CN^{-\frac{1}{2} + \frac{1}{2p}},$$

and B is the integral given by

$$(2.6.19) \qquad B^q = \int_a^x (t - N^{\frac{1}{2}})^{-q/4} e^{-\varepsilon q N^{\frac{1}{4}}(t-N^{\frac{1}{2}})^{3/2}} dt$$

where q is the index conjugate to p. By a change of variables, B gives the estimate

$$(2.6.20) \qquad B \leq C \left(\int_0^\infty t^{-q/4} e^{-dN^{\frac{1}{4}} t^{3/2}} dt \right)^{\frac{1}{q}} \leq C N^{1/24 - 1/6q}.$$

Together they give the estimate $I \leq CN^{-\frac{1}{4}} N^{-\frac{1}{2}+\frac{2}{3p}}$. Since $\frac{4}{3} < p < 2$ it follows that $I \leq CN^{-\frac{1}{4}}$. The final case, namely when $a = (2N)^{\frac{1}{2}}$ and $b = \infty$ is trivial as $|h_n(t)| \leq Ce^{-dt^2}$ with some constant $d > 0$. This completes the proof of the Lemma. ∎

Using Proposition 2.6.1 it is easy to see that G_k^α is an admissible family of operators in the sense of Stein [SW]. In fact, if f and g are simple functions, it follows that

$$(2.6.21) \qquad \left| \int G_k^\alpha f(z) g(z) dz \right| \leq C(1 + |\tau|)^{\frac{1}{2}} \|f\|_1 \|g\|_1$$

where $\alpha = \sigma + i\tau$, uniformly in σ, $0 \leq \sigma \leq 1$. Therefore we can apply analytic interpolation theorem to the family G_k^α. For the operators $G_k^{i\tau}$ and $G_k^{1+i\tau}$ we now prove the following estimates.

Proposition 2.6.2.

(i) $\|G_k^{i\tau} f\|_\infty \leq C(1 + |\tau|)^{\frac{1}{2}} \|f\|_1$,

(ii) $\|G_k^{1+i\tau} f\|_2 \leq C(1 + |\tau|)^n k^{-n} \|f\|_2$.

Proof: The estimate (i) follows from Proposition 2.6.1. To prove (ii) we use Plancherel theorem for the Weyl transform. Since $G_k^\alpha f = f \times \widetilde{\psi}_k^{n\alpha}$ it follows that

$$(2.6.22) \qquad \|G_k^{1+i\tau} f\|_2^2 = (2\pi)^{-n} \|W(f \times \widetilde{\psi}_k^{n+ni\tau})\|_{HS}^2.$$

If T and S are bounded operators on $L^2(\mathbb{R}^n)$, then we know that $\|TS\|_{HS} A \leq \|T\| \|S\|_{HS}$ where $\|T\|$ is the operator norm of T. Therefore,

$$(2.6.23) \qquad \|G_k^{1+i\tau} f\|_2 \leq C \|T_k\| \|W(f)\|_{HS} \leq C \|T_k\| \|f\|_2$$

where T_k is the operator $W(\widetilde{\psi}_k^{n+ni\tau})$. So, it remains to show that $\|T_k\| \leq C(1+|\tau|)^n k^{-n}$.

Using the formula (1.1.38) we can write

$$(2.6.24) \qquad \widetilde{\psi}_k^{n+ni\tau}(z) = \frac{\Gamma(k+1)\Gamma(n+1+ni\tau)}{\Gamma(k+n+1+ni\tau)} \sum_{j=0}^{k} A_{k-j}^{ni\tau} \varphi_j(z).$$

As $W(\varphi_j) = (2\pi)^n P_j$, it follows that

$$(2.6.25) \qquad T_k = \frac{\Gamma(k+1)\Gamma(n+1+ni\tau)}{\Gamma(k+n+1+ni\tau)} \sum_{j=0}^{k} A_{k-j}^{ni\tau} P_j.$$

Therefore, the operator norm of T_k satisfies

$$(2.6.26) \qquad \|T_k\| \leq \max_{1 \leq j \leq k} |A_j^{ni\tau}| \, |A_k^{n+ni\tau}|^{-1}.$$

Using Stirling's formula to estimate the gamma functions it is not difficult to see that $\|T_k\| \leq C(1+|\tau|)^n k^{-n}$. This proves the proposition. ∎

We can now complete the proof of the restriction theorem. Applying analytic interpolation to G_k^α and making use of the preceding proposition we have the estimate

$$(2.6.27) \qquad \|G_k^{(n-1)/n} f\|_{2n/n-1} \leq C k^{-(n-1)} \|f\|_{2n/n+1}.$$

In view of (2.6.4) and the estimate $\frac{\Gamma(k+1)}{\Gamma(k+n)} = O(k^{-n+1})$ it follows that

$$(2.6.28) \qquad \|f \times \varphi_k\|_{2n/n-1} \leq C \|f\|_{2n/n+1}.$$

We also have the estimate $\|f \times \varphi_k\|_\infty \leq \|\varphi_k\|_\infty \|f\|_1 \leq C k^{n-1} \|f\|_1$. Riesz–Thorin convexity theorem proves that

$$(2.6.29) \qquad \|f \times \varphi_k\|_{p'} \leq C k^{2n(\frac{1}{p}-\frac{1}{2})-1} \|f\|_p, \ 1 \leq p \leq \frac{2n}{n+1}.$$

As Q_k are projections it now follows that

$$(2.6.30) \qquad \|Q_k f\|_2^2 = (Q_k f, f) \leq \|f\|_p \|Q_k f\|_{p'} \leq C k^{2n(\frac{1}{p}-\frac{1}{2})-1} \|f\|_p^2.$$

This completes the proof of the restriction theorem. ∎

CHAPTER 3
MULTIPLE HERMITE EXPANSIONS

Our aim in this chapter is to study the almost everywhere and mean convergence of Hermite expansions for functions in $L^p(\mathbb{R}^n)$, $n \geq 2$. It turns out that the series fails to converge in the norm unless $p = 2$. We therefore look at the Riesz means $S_R^\alpha f$ for the Hermite expansions. The main result of this chaper can be stated as the norm convergence of $S_R^\alpha f$ to f for all f in $L^p(\mathbb{R}^n)$, $1 \leq p < \infty$ when $\alpha > (n-1)/2$. It is also proved that there is almost everywhere convergence for $p \geq 2$ under the same condition that $\alpha > (n-1)/2$.

3.1 Riesz Means and the Critical Index

Given a function f on $\mathbb{R}^n$ we have the associated Hermite expansion

$$(3.1.1) \qquad f(x) = \sum_\mu \hat{f}(\mu)\Phi_\mu(x) = \sum_{k=0}^\infty P_k f(x).$$

We know that when f is in $L^2(\mathbb{R}^n)$ the series converges to f in the norm. For f in $L^p(\mathbb{R}^n)$, $p \neq 2$ we like to know if the same thing is true of the series. When $n = 1$ the following fundamental result is known. Let $S_N f$ denote the partial sum

$$(3.1.2) \qquad S_N f(x) = \sum_{k=0}^N P_k f(x).$$

Theorem 3.1.1. *(Askey–Wainger) Assume $n = 1$. The partial sums $S_N f$ converge to f in the L^p norm iff $\frac{4}{3} < p < 4$.*

This theorem is proved in [AW]. We will not attempt a proof of this theorem as it involves estimating a large number (36 to be precise) of integrals. Instead, in Chapter 5 we investigate why the condition $\frac{4}{3} < p < 4$ is necessary for $S_N f$ to converge to f. We postpone the study of one dimensional Hermite series to Chapter 5 as it is more technical. For the rest of this chapter we therefore assume $n \geq 2$. The behavior of the partial sums in this case is quite different. We will show in a moment that the partial sums $S_N f$ never converges to f unless $p = 2$. This follows as we shall show below, from a celebrated theorem of Fefferman on the multiplier problem for the ball through a transplantation technique. Before describing the transplantation some discussion on the multiplier problem is in order.

Consider now the Fourier transform defined for functions on $\mathbb{R}^n$ by

$$(3.1.3) \qquad \widehat{f}(\xi) = (2\pi)^{-\frac{n}{2}} \int_{\mathbb{R}^n} e^{-ix\cdot\xi} f(x)dx.$$

Using the Fourier transform we define the operators $(1+\Delta)_+^\alpha$ by the formula

$$(3.1.4) \qquad (1+\Delta)_+^\alpha f(x) = (2\pi)^{-\frac{n}{2}} \int_{\mathbb{R}^n} e^{ix\cdot\xi} (1-|\xi|^2)_+^\alpha \widehat{f}(\xi)d\xi$$

where $(1-|\xi|^2)_+^\alpha$ stands for the positive part of the function $(1-|\xi|^2)^\alpha$. When $\alpha = 0$, $(1-|\xi|^2)_+^\alpha$ is interpreted as the characteristic function of the unit ball $|\xi| \leq 1$. These operators are related to the summability of n-dimensional Fourier series. For a long time it was known that the inequality

$$(3.1.5) \qquad \|(1+\Delta)_+^\alpha f\|_p \leq C\|f\|_p$$

is valid for all radial functions f in $L^p(\mathbb{R}^n)$ if and only if

$$\frac{2n}{n+1+2\alpha} < p < \frac{2n}{n-1-2\alpha}.$$

It was therefore conjectured that (3.1.5) is true for all functions in the above range of p. The genius of Fefferman disproved the conjecture when $\alpha = 0$ by giving a counterexample: when $\alpha = 0$ (3.1.5) is valid only if $p = 2$. For $\alpha > 0$ the conjecture is proved to be correct for $n = 2$, and for $n > 2$ it has been proved only for $\alpha > \frac{(n-1)}{2(n+1)}$. So much for the multiplier problem. See [Fe] for discussions on the multiplier problem.

We now show that the partial sums $S_N f$ of the Hermite series converge to f in the norm only if $p = 2$.

Theorem 3.1.2. *If the partial sums S_N are uniformly bounded on $L^p(\mathbb{R}^n)$, then $(1+\Delta)_+^\alpha$ with $\alpha = 0$ is bounded on $L^p(\mathbb{R}^n)$. Consequently, $S_N f$ does not converge to f unless $p = 2$.*

Proof: Let $S_N(x,y)$ denote the kernel of S_N which is given by

$$(3.1.6) \qquad S_N(x,y) = \sum_{k=0}^{N} \Phi_k(x,y).$$

We need the following asymptotic behavior of this kernel as $N \to \infty$:

(3.1.7) $$S_N(x,y) = (2\pi)^{-\frac{n}{2}} \int_{|x|^2+|\xi|^2 \leq N} e^{i(x-y)\cdot\xi} d\xi + o(N^{\frac{n}{2}})$$

where the estimate is uniform for x and y in a compact subset of $\mathbb{R}^n$. This result is due to Lars Garding [G]; actually a similar result is true for more general elliptic differential operators.

We observe that when $\alpha = 0$ the operator $(1+\Delta)_+^\alpha$ is given by convolution with a function J. This J is defined by

(3.1.8) $$J(x) = (2\pi)^{-\frac{n}{2}} \int_{|\xi|\leq 1} e^{ix\cdot\xi} d\xi.$$

Let S stand for the operator $(1+\Delta)_+^\alpha$ when $\alpha = 0$ so that

(3.1.9) $$Sf(x) = \int_{\mathbb{R}^n} J(x-y)f(y)dy.$$

Let f and g be in $C_0^\infty(\mathbb{R}^n)$. To prove the theorem it is therefore enough to show

(3.1.10) $$\left|\int Sf(x)g(x)dx\right| \leq C\|f\|_p\|g\|_{p'}.$$

Define $f_R(x) = f(Rx)$, $g_R(x) = g(Rx)$. Since the partial sums S_{R^2} are assumed to be uniformly bounded on $L^p(\mathbb{R}^n)$ we have

(3.1.11) $$\left|\int_{\mathbb{R}^n} S_{R^2}f_R(x)g_R(x)dx\right| \leq C\|f_R\|_p\|g_R\|_{p'}.$$

After a change of variables this gives

(3.1.12) $$R^{-n}\left|\int_{\mathbb{R}^n}\int_{\mathbb{R}^n} S_{R^2}(x/R,y/R)f(y)g(x)dydx\right| \leq C\|f\|_p\|g\|_{p'}.$$

To complete the proof it is therefore enough to show that

(3.1.13) $$\lim_{R\to\infty} R^{-n} \int_{\mathbb{R}^n}\int_{\mathbb{R}^n} S_{R^2}(x/R,y/R)f(y)g(x)dxdy$$
$$= \int_{\mathbb{R}^n}\int_{\mathbb{R}^n} J(x-y)f(y)g(x)dx.$$

Since f and g are compactly supported, it suffices to show that

(3.1.14) $$\lim_{R\to\infty} R^n S_{R^2}(x/R, y/R) = J(x-y).$$

This follows immediately from a change of variables in the asymptotic formula (3.1.7). This completes the proof of the theorem. ∎

As the partial sums fail to converge we are led to consider Riesz means. These are defined by

(3.1.15) $$S_R^\alpha f = \sum_{k=0}^\infty \left(1 - \frac{2k+n}{R}\right)_+^\alpha P_k f.$$

We can also consider Cesaro means as in the one dimensional case. As we have already remarked in Chapter 2, we are considering Riesz means here as it is suitable for the method we use to study. Regarding S_R^α we have the following conjecture to make: for $\alpha > 0$, the uniform estimates

(3.1.16) $$\|S_R^\alpha f\|_p \leq C\|f\|_p$$

hold for f in $L^p(\mathbb{R}^n)$ if and only if p lies in the interval $2n/(n+1+2\alpha) < p < 2n/(n-1-2\alpha)$.

It is possible to use a transplantation argument as in the case of the previous theorem to show that the conditions on p are necessary for (3.1.16) to be true. Instead of attempting a proof now, we would rather wait until we reach Section 3.4 where the above conjecture is proved for all radial functions. In fact, the above conjecture has been proved only for radial functions. In the general case we only know that the conjecture is true for $\alpha > (n-1)/2$. From the above conjecture, it is clear that S_R^α cannot be uniformly bounded on $L^1(\mathbb{R}^n)$ unless $\alpha > (n-1)/2$. Thus $\alpha = (n-1)/2$ is the critical index for the Riesz summability.

The next two sections are devoted to proving summability of the Hermite series above the critical index. We conclude this section with some remarks regarding the method we use to study the Riesz means. It is possible to express the Riesz kernel as a sum of certain oscillatory integrals as we will do for the one dimensional case in Chapter 5 and then try to estimate the oscillatory integrals. But that procedure turns out to be rather clumsy. Instead, we obtain, following a method of Peetre [P], an L^p estimate for the Reisz kernel. However, this method works only when $n \geq 2$.

3.2 An L^2 Estimate for the Riesz Kernel

Consider the kernel of the Riesz means

$$(3.2.1) \qquad S_R^\alpha(x,y) = \sum_{k=0}^{\infty} \left(1 - \frac{2k+n}{R}\right)_+^\alpha \Phi_k(x,y).$$

Our aim in this section is to get an estimate for the L^2 norm of $S_R^\alpha(x,y)$ when $|x - y| > r$. In order to obtain the estimate we need several preliminary estimates and formulas for certain kernels defined by Hermite functions. We begin with a very simple estimate for the kernel of the partial sum operator S_N. Recall that the kernel $S_N(x,y)$ of S_N is given by

$$(3.2.2) \qquad S_N(x,y) = \sum_{k=0}^{N} \Phi_k(x,y).$$

Lemma 3.2.1. *There is a constant C independent of N, x and y such that for all $x, y \in \mathbb{R}^n$*

$$(3.2.3) \qquad |S_N(x,y)| \le CN^{\frac{n}{2}}.$$

Proof: The proof is based on Mehler's formula,

$$(3.2.4) \qquad \sum_{k=0}^{\infty} \Phi_k(x,y) r^k = \pi^{-\frac{n}{2}}(1-r^2)^{-\frac{n}{2}} e^{-\frac{1}{2}\frac{1+r^2}{1-r^2}(|x|^2+|y|^2) + \frac{2rx \cdot y}{1-r^2}},$$

from which it follows for any $0 < r < 1$,

$$(3.2.5) \qquad r^N \sum_{k=0}^{N} \Phi_k(x,x) r^k \le \sum_{k=0}^{\infty} \Phi_k(x,x) r^k$$

$$= \pi^{-\frac{n}{2}}(1-r^2)^{-\frac{n}{2}} e^{-\frac{1}{2}\frac{1-r}{1+r}|x|^2} \le C(1-r)^{-\frac{n}{2}}.$$

Taking $r = e^{-\frac{1}{N}}$ we complete the proof. ∎

The next lemma is about the kernel $\Phi_k(x,y)$ itself. In view of Lemma 3.2.1 it is reasonable to conjecture that $|\Phi_N(x,x)| \le CN^{\frac{n}{2}-1}$; the next lemma shows that this is indeed true.

Lemma 3.2.2. *Asume that $n \geq 2$. Then there is a constant C independent of N and x such that*

(3.2.6) $$|\Phi_N(x,x)| \leq CN^{n/2-1}.$$

We remark in passing that this lemma is not true when $n = 1$. In fact, in that case $\Phi_N(x,x) = (h_N(x))^2$ and the L^∞ norm of h_N behaves like $N^{-\frac{1}{12}}$.

Proof: We prove the lemma by induction. Assume that it is true for $n = m$, $m \geq 2$ and consider the case $n = m+1$. Writing $x = (y,t)$, $y \in \mathbb{R}^m$, $t \in \mathbb{R}$, we have

(3.2.7) $$\Phi_N(x,x) = \sum_{|\mu|+j=N} (\Phi_\mu(y))^2 (h_j(t))^2$$

$$= \sum_{j=0}^{N} (h_j(t))^2 \Phi_{N-j}(y,y)$$

$$\leq C \sum_{j=0}^{N} (N-j)^{\frac{m}{2}-1} (h_j(t))^2$$

by the induction hypothesis. By applying Lemma 3.2.1 with $n = 1$ the above gives the estimate

(3.2.8) $$\Phi_N(x,x) \leq CN^{\frac{m}{2}-1} \sum_{j=0}^{N} (h_j(t))^2 \leq CN^{(m+1)/2-1}.$$

So, it suffices to prove the lemma when $n = 2$. Writing $x = (t,s)$ we have

$$\Phi_N(x,x) = \sum_{j=0}^{N} (h_j(t))^2 (h_{N-j}(s))^2.$$

In view of the Mehler's formula (3.2.4) it is clear that $\Phi_N(x,x)$ is a radial function. Therefore, if $r = |x|$ then we have

(3.2.9) $$\Phi_N(x,x) = \sum_{j=0}^{N} (h_j(r))^2 (h_{N-j}(0))^2.$$

Now we know that $h_{2j+1}(0) = 0$ and

(3.2.10) $$(h_{2j}(0))^2 = \pi^{-1} \frac{\Gamma(j+\frac{1}{2})}{\Gamma(j+1)}.$$

This has been proved in (1.1.21). Assuming that $N = 2m$ is even we have

$$(3.2.11) \qquad \Phi_N(x,x) = \frac{1}{\pi} \sum_{j=0}^{m} (h_{2j}(r))^2 \frac{\Gamma(m-j+\frac{1}{2})}{\Gamma(m-j+1)}.$$

Using Stirling's formula for the gamma function we have

$$(3.2.12) \qquad \Phi_N(x,x) \leq C \sum_{j=0}^{m} (h_{2j}(r))^2 (N - 2j + 1)^{-\frac{1}{2}}.$$

Next we make use of the following estimates for the Hermite functions (see Lemma 1.5.1):

$$(3.2.13) \qquad |h_j(\gamma)| \leq C(1 + 2j + 1 - r^2)^{-\frac{1}{4}}, \quad \text{if } r^2 \leq 2j + 1,$$
$$\leq Ce^{-dr^2}, \qquad \qquad \text{if } r^2 > 2j + 1$$

where $d > 0$ is a fixed constant.

We split the sum in (3.2.12) into two parts. Let j_0 be the smallest integer such that $r^2 \leq 2j_0 + 1$. The sum

$$(3.2.14) \qquad \sum_{j=0}^{j_0-1} (h_{2j}(r))^2 (N - 2j + 1)^{-\frac{1}{2}}$$

is clearly bounded since $|h_{2j}(r)|^2 \leq Ce^{-2d\gamma^2}$. For the other sum we have

$$\sum_{j=j_0}^{m} (h_{2j}(r))^2 (N - 2j + 1)^{-\frac{1}{2}} \leq C \sum_{j=j_0}^{m} (1 + J - r^2)^{-\frac{1}{2}} (N + 2 - J)^{-\frac{1}{2}}$$

where $J = 2j + 1$. Since $r^2 \leq 2j_0 + 1 = J_0$, the above sum is bounded by

$$\sum_{j=j_0}^{m} (1 + j - J_0)^{-\frac{1}{2}} (N + 2 - J)^{-\frac{1}{2}} \leq \sum_{j=J_0}^{2m+1} (1 + j - J_0)^{-\frac{1}{2}} (N + 2 - j)^{-\frac{1}{2}}$$
$$\leq \sum_{j=0}^{a} (1 + j)^{-\frac{1}{2}} (a + 1 - j)^{-\frac{1}{2}}$$

where $a = 2m + 1 - J_0$. The last sum is clearly bounded. This proves that $\Phi_N(x,x) \leq C$ when $N = 2m$. The case $N = 2m + 1$ is similar. This proves the lemma. ∎

We need one more lemma before we could proceed to get an L^2 estimate for the Riesz kernel $S_R^\alpha(x,y)$. The following lemma gives a precise expression for $(x-y)^\beta S_R^\alpha(x,y)$ where β is a multiindex. More generally, given a function ψ defined on the set of non-negative integers we consider the kernel

$$(3.2.15) \qquad M_\psi(x,y) = \sum_{N=0}^{\infty} \psi(N)\Phi_N(x,y)$$

and find a formula for $(x-y)^\beta M_\psi(x,y)$. To state the lemma we need to introduce some notation. Let Δ be the finite difference operator defined by

$$(3.2.16) \qquad \Delta\psi(N) = \psi(N+1) - \psi(N)$$

and let Δ^k be defined inductively. We set $A_j = \left(-\frac{\partial}{\partial x_j} + x_j\right)$ and $B_j = \left(-\frac{\partial}{\partial y_j} + y_j\right)$ to be the creation operators and define for any multiindex γ the operator $(B-A)^\gamma = \prod_{j=1}^{n}(B_j - A_j)^{\gamma_j}$. By $\Delta^k M_\psi(x,y)$ we denote the kernel $M_{\Delta^k \psi}(x,y)$. With this notation we have the following.

Lemma 3.2.3. *For every multiindex β we can write*

$$(x-y)^\beta M_\psi(x,y) = \sum C_{\gamma\delta}(B-A)^\gamma \Delta^{|\delta|} M_\psi(x,y),$$

where the sum is extended over all multiindices γ and δ satisfying

$$2\delta_j - \gamma_j = \beta_j, \quad \delta_j \leq \beta_j.$$

Proof: We recall the following two formulas for the Hermite functions h_k from Chapter 1.

$$(3.2.17) \qquad \left(-\frac{d}{dt}+t\right)h_k(t) = (2k+2)^{\frac{1}{2}} h_{k+1}(t),$$

$$(3.2.17') \qquad 2th_k(t) = (2k+2)^{\frac{1}{2}} h_{k+1}(t) + (2k)^{\frac{1}{2}} h_{k-1}(t).$$

Recalling the definition of $\Phi_\mu(x)$, we have the formulas

$$(3.2.18) \qquad 2x_j \Phi_\mu(x) = (2\mu_j + 2)^{\frac{1}{2}} \Phi_{\mu+e_j}(x) + (2\mu_j)^{\frac{1}{2}} \Phi_{\mu-e_j}(x),$$

$$(3.2.19) \qquad A_j \Phi_\mu(x) = (2\mu_j + 2)^{\frac{1}{2}} \Phi_{\mu+e_j} \quad \text{and}$$

$$B_j \Phi_\mu(x) = (2\mu_j + 2)^{\frac{1}{2}} \Phi_{\mu+e_j}(y).$$

These relations are very important for the following calculations.

We now calculate $2(x_j - y_j)M_\psi(x,y)$. Using the recursion relation (3.2.18) and rearranging we see that

(3.2.20) $$2(x_j - y_j)M_\psi(x,y) = (B_j - A_j)\Delta M_\psi(x,y)$$

where we have made use of (3.2.19) also. We claim that iteration produces

(3.2.21) $$2^m(x_j - y_j)^m M_\psi(x,y) = \sum C_{rs}(B_j - A_j)^r \Delta^s M_\psi(x,y)$$

where the sum is extended over all r and s satisfying $2s - r = m$, $s \leq m$. We prove this claim by induction. Assuming that (3.2.21) is true for $m = k$ we consider the case $m = (k+1)$. We have

(3.2.22) $$2^{k+1}(x_j - y_j)^{k+1} M_\psi(x,y)$$
$$= \sum C_{rs} 2(x_j - y_j)(B_j - A_j)^r \Delta^s M_\psi(x,y)$$

with $2s - r = k$, $s \leq k$ by the induction hypothesis. If we know that

(3.2.23) $$(x_j - y_j)(B_j - A_j)^r - (B_j - A_j)^r(x_j - y_j) = -2r(B_j - A_j)^{r-1}$$

then we have

(3.2.24) $$2^{k+1}(x_j - y_j)^{k+1} M_\psi(x,y)$$
$$= \sum C_{rs}\Big\{-2r(B_j - A_j)^{r-1} + (B_j - A_j)^r 2(x_j - y_j)\Big\}\Delta^s M_\psi(x,y)$$

which can be written in the form

$$\sum C'_{r's'}(B_j - A_j)^{r'} \Delta^{s'} M_\psi(x,y)$$

with r' and s' satisfying $2s' - r' = k+1$, $s' \leq k+1$. Thus the claim is proved modulo (3.2.23).

Letting $T = (x_j - y_j)$ and $S = (B_j - A_j)$ an easy calculation shows that the commutator $[T, S] = -2I$. In order to prove (3.2.23) it is enough to show that $[T, S^r] = -2rS^{r-1}$. Again we prove this by induction. Assuming $[T, S^r] = -2rS^{r-1}$ we consider

$$[T, S^{r+1}] = (TS^r - S^r T + S^r T)S - S(S^r T - TS^r + TS^r)$$
$$= -4rS^r + S^r(TS - ST + ST) - (ST - TS + TS)S^r$$
$$= -4(r+1)S^r - [T, S^{r+1}].$$

This proves that $[T, S^{r+1}] = -2(r+1)S^r$.

Now the operators $(x_i - y_i)$ and $(B_j - A_j)$ commute with each other when $i \neq j$. Therefore, from (3.2.22) it follows that

(3.2.25) $\quad (x - y)^\beta M_\psi(x,y) = \sum C_{\gamma\delta}(B - A)^\gamma \Delta^{|\delta|} M_\psi(x,y)$

with $2\delta_j - \gamma_j = \beta_j$, $\delta_j \leq \beta_j$. This completes the proof of the lemma. ∎

Now we can prove the following estimate for the kernel of the Riesz means.

Theorem 3.2.1. *Assume that $n \geq 2$. Then*

$$\left(\int_{|x-y|\geq r} |S_R^\alpha(x,y)|^2 dy\right)^{\frac{1}{2}} \leq CR^{\frac{n}{4}}(1 + R^{\frac{1}{2}}r)^{-\alpha-\frac{1}{2}}$$

where $C > 0$ is independent of R and x.

Proof: In view of the orthonormality of the Hermite functions Φ_μ the square of the L^2 norm of the kernel $S_R^\alpha(x,y)$ equals

(3.2.26) $\quad \sum (1 - \nu/R)_+^\alpha \Phi_N(x,x)$

where we have set $\nu = 2N + n$. In view of Lemma 3.2.1 the above is bounded by $CR^{\frac{n}{2}}$, and so the estimate of the theorem is valid when $R^{\frac{1}{2}}r \leq 1$. It is therefore enough to prove the theorem when $R^{\frac{1}{2}}r > 1$.

The idea of the proof is to split the kernel into two parts and estimate them separately. To do the splitting we take a real number t, $0 < t < 1$ and choose a C^∞ function ω_t such that $\omega_t(s) = 1$ for $s < 1-t$, $\omega_t(s) = 0$ for $s > 1 - t/2$ and $|\omega_t^{(k)}(s)| \leq C_k t^{-k}$ on $1 - t < s < 1 - t/2$. With this choice of ω_t we define

(3.2.27) $\quad S_{R,1}^\alpha(x,y) = \sum (1 - \nu/R)_+^\alpha (1 - \omega_t(\nu/R))\Phi_N(x,y),$

(3.2.28) $\quad S_{R,2}^\alpha(x,y) = \sum (1 - \nu/R)_+^\alpha \omega_t(\nu/R)\Phi_N(x,y).$

For the kernel $S_{R,1}^\alpha$ we have the following:

Lemma 3.2.4. *There is a constant C such that*

$$\int_{\mathbb{R}^n} |S_{R,1}^\alpha(x,y)|^2 dy \leq CR^{n/2}t^{2\alpha+1}.$$

Proof: The square of the L^2 norm of the kernel $S_{R,1}^\alpha(x,y)$ is given by

$$\sum (1-\nu/R)_+^{2\alpha}(1-\omega_t(\nu/R))^2 \Phi_N(x,x).$$

Since $1-t < \nu/R$ on the support of $(1-\omega_t(\nu/R))$, we have

$$\int_{\mathbb{R}^n} |S_{R,1}^\alpha(x,y)|^2 dy \leq Ct^{2\alpha} \sum_{1-t\leq \nu/R \leq 1} \Phi_N(x,x).$$

Using the estimate $\Phi_N(x,x) \leq CN^{\frac{n}{2}-1}$ we obtain

$$\int_{\mathbb{R}^n} |S_{R,1}^\alpha(x,y)|^2 dy \leq CR^{\frac{n}{2}-1}t^{2\alpha}(Rt).$$

This proves the lemma. ∎

Next we take the kernel $S_{R,2}^\alpha(x,y)$ and prove the following lemma.

Lemma 3.2.5. *Assume that m is an integer greater than $2\alpha + 1$. Then there is a constant C such that*

$$\int_{\mathbb{R}^n} |x-y|^{2m} |S_{R,2}^\alpha(x,y)|^2 dy \leq CR^{-m+n/2} t^{2\alpha+1-2m}.$$

Proof: It is enough to prove the estimate

(3.2.29) $$\int_{\mathbb{R}^n} |(x-y)^\beta S_{R,2}^\alpha(x,y)|^2 dy \leq CR^{-m+n/2} t^{2\alpha+1-2m},$$

for every β with $|\beta| = m$. If we set $\psi(N) = (1-\nu/R)_+^\alpha \omega_t(\nu/R)$ so that $S_{R,2}^\alpha(x,y) = M_\psi(x,y)$, then in view of Lemma 2.2.3 we have

(3.2.30) $$(x-y)^\beta S_{R,2}^\alpha(x,y) = \sum C_{\gamma,k}(B-A)^\gamma \Delta^k M_\psi(x,y)$$

where $2k - |\gamma| = m$, $k \leq m$. On expanding $(B-A)^\gamma$ a typical term of the above sum is

(3.2.31) $$\sum \Delta^k \psi(|\mu|) A^\tau \Phi_\mu(x) B^\sigma \Phi_\mu(y)$$

where $2k - |\sigma| - |\tau| = m$, $k \leq m$. In view of the effect of A and B on Hermite functions the square of the L^2 norm of the above sum is bounded by a constant times

(3.2.32) $$\sum_{N=0}^\infty |\Delta^k \psi(N)|^2 (2N+n)^{|\sigma|+|\tau|} \Phi_{N+|\tau|}(x,x).$$

If we use the estimate $\Phi_N(x,x) \leq CN^{n/2-1}$ and the relation $|\sigma|+|\tau| = 2k - m$ the above is bounded by

$$(3.2.33) \qquad \sum_{N=0}^{\infty} |\Delta^k \psi(N)|^2 (2N+n)^{2k-m+n/2-1}.$$

We shall show that this sum is bounded by a constant times $R^{-m+n/2} t^{2\alpha+1-2m}$.

We express the finite differences in terms of derivatives as follows. We have

$$\Delta \psi(N) = \int_0^1 \psi'(N+s_1)ds_1,$$

which gives by iteration

$$\Delta^k \psi(N) = \int_0^1 \cdots \int_0^1 \psi^{(k)}(N+s_1+\cdots+s_k)ds_1 \cdots ds_k.$$

Therefore,

$$|\Delta^k \psi(N)|^2 \leq \int_0^1 \cdots \int_0^1 |\psi^{(k)}(N+s_1+\cdots+s_k)|^2 ds_1 \cdots ds_k$$

and hence it is enough to prove that

$$(3.2.34) \qquad \sum_{N=0}^{\infty} |\psi^{(k)}(N+s)|^2 (2N+m)^{2k-m+n/2-1}$$
$$\leq CR^{-m+n/2} t^{2\alpha+1-2m}$$

with a constant C independent of s, $0 \leq s \leq m$. Since

$$\psi(N) = (1-\nu/R)_+^\alpha \, \omega_t(\nu/R)$$

it follows that

$$(3.2.35) \qquad \psi^{(k)}(N) = R^{-k} 2^k \sum_{j=0}^{k} a_j (1-\nu/R)_+^{\alpha-j} \omega_t^{(k-j)}(\nu/R).$$

So we will estimate a typical term

$$(3.2.36) \qquad R^{-2k} \sum_{N=0}^{\infty} \left(1 - \frac{\nu+2s}{R}\right)_+^{2\alpha-2j}$$
$$\times \left|\omega_t^{(k-j)}\left(\frac{\nu+2s}{R}\right)\right|^2 (\nu+2s)^{2k-m+\frac{n}{2}-1}.$$

We treat two cases. When $k - j \neq 0$, the functions $\omega_t^{(k-j)}\left(\frac{\nu+2s}{R}\right)$ are supported in the interval $1 - t < \frac{\nu+2s}{R} < 1 - t/2$. Since $|\omega_t^{(k-j)}\left(\frac{\nu+2}{R}\right)| \leq C_j t^{-k+j}$ we see that the above sum (3.2.36) is bounded by

$$(3.2.37) \quad R^{-2k}t^{-2k+2\alpha} \sum_{1-t \leq \frac{\nu+2s}{R} \leq 1-t/2} (\nu+2s)^{2k-m+n/2-1}$$

which in turn is bounded by $R^{-m+n/2-1}t^{-2k+2\alpha}(Rt)$. As $2\alpha + 1 - 2k > 2\alpha + 1 - 2m$ we get the bound $CR^{-m+n/2}t^{2\alpha+1-2m}$.

Next we consider the case $j = k$. In this case the sum (3.2.36) is extended over the interval $\nu + 2s \leq R(1 - t/2)$ and is bounded by

$$(3.2.38) \quad R^{-m+n/2-1} \sum_{\nu+2s \leq R(1-t/2)} \left(1 - \frac{\nu+2s}{R}\right)_+^{2\alpha-2k}.$$

The sum taken over $R(1 - t) \leq \nu + 2s \leq R(1 - t/2)$ gives the same estimate as before. The remaining part of the sum is bounded by

$$(3.2.39) \quad R^{-m+n/2} \int_0^{1-t} (1-\lambda)^{2\alpha-2k} d\lambda = R^{-m+n/2} \int_t^1 \lambda^{2\alpha-2k} d\lambda.$$

Since $2k - |\gamma| = m$, $2k \geq m$ or $2\alpha - 2k \leq 2\alpha - m$ so that $2\alpha - 2k + 1 \leq 2\alpha + 1 - m < 0$ by the choice of m. Therefore, the above integral is bounded by $Ct^{2\alpha+1-2k} \leq Ct^{2\alpha+1-2m}$. This completes the proof of Lemma 3.2.5. ∎

We now complete the proof of Theorem 3.2.1. When $|x - y| \geq r$, writing

$$|S_R^\alpha(x,y)|^2 \leq 2|S_{R,1}^\alpha(x,y)|^2 + 2r^{-2m}|x-y|^{2m}|S_{R,2}^\alpha(x,y)|^2$$

we obtain, from the two lemmas,

$$\int_{|x-y| \geq r} |S_R^\alpha(x,y)|^2 dy \leq C\{R^{n/2}t^{2\alpha+1} + r^{-2m}R^{-m+n/2}t^{2\alpha+1-2m}\}.$$

The choice $t = (R^{1/2}r)^{-1}$ proves the theorem. ∎

3.3 Almost Everywhere and Mean Summability

In this section we have results concerning the almost everywhere and mean convergence of the Riesz means. From the L^2 estimate of the Riesz kernel we deduce an L^p estimate which will be used in the proof of the norm convergence.

Theorem 3.3.1. If $1 \le p \le 2$ and $\alpha > (n-1)/2$ then

$$\left(\int_{|x-y| \ge r} |S_R^\alpha(x,y)|^p dy \right)^{\frac{1}{r}} \le C R^{n/2q} (1 + R^{\frac{1}{2}} r)^{-\alpha - \frac{1}{2} + n(\frac{1}{p} - \frac{1}{2})}$$

where $\frac{1}{p} + \frac{1}{q} = 1$ and C is independent of R, r and x.

Proof: Consider a partition of $\mathbb{R}^n$ into dyadic annuli $A_i = \{y : 2^i r \le |x-y| \le 2^{i+1} r\}$. Applying Holder's inequality we obtain

(3.3.1)
$$\left(\int_{|x-y| \ge r} |S_R^\alpha(x,y)|^p dy \right)^{\frac{1}{p}}$$
$$\le \sum_{i=0}^\infty |A_i|^{(2-p)/2p} \left(\int_{A_i} |S_R^\alpha(x,y)|^2 dy \right)^{\frac{1}{2}}.$$

Since the Lebesgue measure of A_i, $|A_i| \le C(2^i r)^n$, using the estimate of Theorem 3.2.1 we get

(3.3.2)
$$\left(\int_{|x-y| \ge r} |S_R^\alpha(x,y)|^p dy \right)^{\frac{1}{p}} \le C \sum_{i=0}^\infty \frac{R^{n/4}(2^i r)^{n(\frac{1}{p} - \frac{1}{2})}}{(1 + R^{\frac{1}{2}} 2^i r)^{\alpha + \frac{1}{2}}}.$$

The sum on the right hand side of (3.3.2) is equal to

(3.3.3)
$$R^{n/2q} \sum_{i=0}^\infty (2^i t)^a (1 + 2^i t)^{-b} = R^{n/2q} F(t)$$

where $t = R^{\frac{1}{2}} r$, $a = n(\frac{1}{p} - \frac{1}{2})$ and $b = \alpha + \frac{1}{2}$. Once we know that $F(t) \le C(1+t)^{a-b}$ we are done. The estimate for $F(t)$ is clearly valid for $t \ge 1$ as $0 \le a < b$, and for $0 < t < 1$ $F(t)$ is bounded by

(3.3.4)
$$G(t) = \sum_{i=-\infty}^\infty (2^i t)^a (1 + 2^i t)^{-b}$$

which in turn being bounded on $[1,2]$ and satisfying $G(2^i t) = G(t)$, is also bounded on $(0, \infty)$. This proves the theorem. ∎

Now we have the following summability result.

Theorem 3.3.2. Let $1 \leq p \leq \infty$, $f \in L^p(\mathbb{R}^n)$ and $\alpha > \frac{(n-1)}{2}$. Then we have the uniform estimates

$$\|S_R^\alpha f\|_p \leq C\|f\|_p.$$

Moreover, $S_R^\alpha f$ converges to f in the norm as $R \to \infty$ for $f \in L^p(\mathbb{R}^n)$, $1 \leq p < \infty$.

Proof: Letting $r \to 0$ in the previous theorem we see that $S_R^\alpha(x, y)$ are uniformly integrable for $\alpha > (n-1)/2$. This immediately proves the uniform boundedness of S_R^α on all $L^p(\mathbb{R}^n)$, $1 \leq p \leq \infty$. The norm convergence can be proved using a density argument (see Lemma 5.4.1). ∎

We now establish a maximal inequality which will give us almost everywhere convergence. Let S_*^α stand for the maximal operator

$$(3.3.5) \qquad S_*^\alpha f(x) = \sup_{R>0} |S_R^\alpha f(x)|.$$

In what follows $M_p f$ stands for the Hardy–Littlewood maximal function of $|f|^p$ raised to the power $\frac{1}{p}$,

$$(3.3.6) \qquad M_p f(x) = \left(\sup_{r>0} \frac{1}{B_r(x)} \int_{B_r(x)} |f(y)|^p \, dy \right)^{\frac{1}{p}}.$$

Here $B_r(x)$ is the ball of radius r centered at x.

Theorem 3.3.3. Assume that $f \in L^p(\mathbb{R}^n)$, $p \geq 2$ and $\alpha > (n-1)/2$. Then we have the pointwise inequality

$$S_*^\alpha f(x) \leq C M_p f(x).$$

Thus S_*^α is bounded on $L^p(\mathbb{R}^n)$, $p \geq 2$ and consequently, $S_R^\alpha f(x)$ converges to $f(x)$ almost everywhere.

Proof: The proof is based on the estimate of Theorem 3.3.1. Let f be a function vanishing in $|x - y| \leq r$. Then

$$(3.3.7) \qquad |S_R^\alpha f(x)| \leq \left(\int_{|x-y| \geq r} |S_R^\alpha(x, y)|^q \, dy \right)^{\frac{1}{q}} \|f\|_p$$

where $\frac{1}{p} + \frac{1}{q} = 1$. Since $q \leq 2$ we have

$$(3.3.8) \qquad \left(\int_{|x-y| \geq r} |S_R^\alpha(x, y)|^q \, dy \right)^{\frac{1}{q}} \leq C R^{n/2p} (1 + R^{\frac{1}{2}} r)^{-\alpha - \frac{1}{2} + n(\frac{1}{q} - \frac{1}{2})}.$$

Given f we set $f_k(y) = f(y)$ for $2^k \leq |x - y| \leq 2^{k+1}$ and $f_k(y) = 0$ otherwise. Then

(3.3.9) $\quad |S_R^\alpha f(x)| \leq \sum_{k=-\infty}^{\infty} |S_R^\alpha f_k(x)|$

$$\leq CR^{n/2p} \sum_{k=-\infty}^{\infty} (1 + R^{\frac{1}{2}}2^k)^{-\alpha - \frac{1}{2} + n(\frac{1}{q} - \frac{1}{2})} \|f_k\|_p.$$

Since $\|f_k\|_p \leq C(2^k)^{n/p} M_p f(x)$ the above sum is dominated by $CM_p f(x)$ times

(3.3.10) $\quad \sum_{k=-\infty}^{\infty} R^{n/2p}(2^k)^{n/p}(1 + R^{\frac{1}{2}}2^k)^{-\alpha - \frac{1}{2} + n(\frac{1}{q} - \frac{1}{2})}$

$$= \sum_{k=-\infty}^{\infty} (R^{\frac{1}{2}}2^k)^{n/p}(1 + R^{\frac{1}{2}}2^k)^{-\alpha - \frac{1}{2} + n(\frac{1}{q} - \frac{1}{2})}$$

$$= G(R^{\frac{1}{2}}).$$

The function

$$G(t) = \sum_{k=-\infty}^{\infty} (2^k t)^{n/p}(1 + 2^k t)^{-\alpha - \frac{1}{2} + n(\frac{1}{q} - \frac{1}{2})}$$

is clearly locally bounded and since $G(2^i t) = G(t)$ it is bounded on $(0, \infty)$. Hence we have $S_*^\alpha f(x) \leq CM_p f(x)$. The almost everywhere convergence now follows in a routine way since $S_R^\alpha f(x)$ converges to $f(x)$ uniformly for C_0^∞ functions. ∎

We have proved the almost everywhere convergence only for f in $L^p(\mathbb{R}^n)$, $p \geq 2$. It is also possible to establish almost everywhere convergence for $1 \leq p \leq 2$. In fact as in the one dimensional case, we can express the Cesàro means for the Hermite expansions as a sum of oscillatory integrals and estimate them using the method of stationary phase. We have remarked earlier that this method turns out to be cumbersome in the higher dimensional case. Nevertheless, we do get an almost everywhere convergence result for $1 \leq p \leq 2$ using that method. We simply state the result without giving a proof.

Theorem 3.3.4. *Assume that $1 \leq p < \infty$ and $\alpha > (3n - 2)/6$. Then for every f in $L^p(\mathbb{R}^n)$, $S_R^\alpha f(x)$ converges to $f(x)$ almost everywhere.*

As in the one dimensional case we can also prove analogues of Fejer-Lebesgue theorem and Riemann's localization principle. All these results follow from the basic estimate stated below.

Theorem 3.3.5. For $\alpha > (3n-2)/6$ the following estimate is valid:

$$|S_R^\alpha(x,y)| \leq CR^{\frac{n}{2}}\left\{(1+R^{\frac{1}{2}}|x-y|)^{-\alpha+(3n-2)/6-1}\right.$$
$$\left. + (1+R^{\frac{1}{2}}|x+y|)^{-\alpha+(3n-2)/6-1}\right\}.$$

We refer to the paper [T3] for details. For $n = 1$ the above two theorems are proved in Chapter 5.

3.4 Hermite Expansions for Radial Functions

In this section we consider Hermite expansions for radial functions. As we have remarked earlier in Section 3.1 we can prove the conjecture that S_R^α are uniformly bounded for radial functions in $L^p(\mathbb{R}^n)$, p verifying $2n/n+1+2\alpha < p < 2n/n-1-2\alpha$. The proof of this result needs two ingredients. First, we need the L^1 estimate of the Riesz kernel proved in the previous section. The second ingredient is the bound

$$(3.4.1) \qquad \|P_k f\|_2 \leq Ck^{\frac{n}{2}(\frac{1}{p}-\frac{1}{2})-\frac{1}{2}}\|f\|_p, \ 1 \leq p < \frac{2n}{n+1}$$

which is called the $L^p - L^2$ restriction theorem. This method of using the $L^p - L^2$ estimates for the projection operators and a kernel estimate to prove uniform boundedness of S_R^α has been explained in detail in Chapter 2 where we studied special Hermite expansions.

The main aim of this section is to prove a Hecke–Bochner type identity for the Hermite projection operators P_k. As a corollary we prove that the Hermite series reduces to a Laguerre series when we consider only radial functions. This corollary together with certain L^p estimates for Laguerre functions will then establish (3.4.1).

To state the main result of this section let us recall the Hecke–Bochner identity for the Fourier transform on $\mathbb{R}^n$. Let $P(x)$ be a solid harmonic of degree m. By this we mean that $P(x)$ is homogeneous of degree m, $P(\lambda x) = \lambda^m P(x)$ and satisfies the Laplace equation, $\Delta P(x) = 0$. Consider a function f on $\mathbb{R}^n$ of the form $f(x) = f_0(|x|)P(x)$. Then the Hecke–Bochner identity for the Fourier transform says that $\hat{f}$ is also of the same form. More precisely, $\hat{f}(\xi) = F_0(|\xi|)P(\xi)$ where

$$(3.4.2) \qquad F_0(r) = (2\pi)(i)^{-m}r^{-(n/2+m-1)}$$
$$\times \int_0^\infty f_0(s)J_{n/2+m-1}(2\pi rs)s^{\frac{n}{2}+m}ds$$

where J_α is the Bessel function of order α. In this section we prove a similar formula for the projections P_k. For a proof of (3.4.2) see [SW].

82   CHAPTER 3

Let L_k^δ be the Laguerre polynomials of type δ defined in Chapter 1. For a function f on $(0, \infty)$ define

$$(3.4.3) \qquad R_k^\delta(f) = 2\frac{\Gamma(k+1)}{\Gamma(k+\delta+1)} \int_0^\infty f(r) L_k^\delta(r^2) e^{-\frac{1}{2}r^2} r^{2\delta+1} dr.$$

We can now state and prove the following theorem.

Theorem 3.4.1. Assume that $f(x) = f_0(|x|) P(x)$ where P is a solid harmonic of degree m. Then one has $P_{2k+m}(f) = F_k(|x|) P(x)$ where

$$(3.4.4) \qquad F_k(r) = R_k^\delta(f_0) L_k^\delta(r^2) e^{-\frac{1}{2}r^2}$$

with $\delta = (\frac{n}{2} + m - 1)$. For other values of j, $P_j f = 0$.

Corollary 3.4.1. If $f(x) = f_0(|x|)$ is radial, then $P_{2k+1}(f) = 0$ and

$$P_{2k}(f) = R_k^{n/2-1}(f_0) L_k^{n/2-1}(r^2) e^{-\frac{1}{2}r^2}.$$

Proof: We start with Mehler's formula for the Hermite functions on $\mathbb{R}^n$. We let $G_\omega(r) = \exp(-\frac{1}{2}\frac{1+\omega^2}{1-\omega^2} r)$ so that for $|\omega| < 1$

$$(3.4.5) \qquad \sum_{k=0}^\infty \Phi_k(x,y) \omega^k$$
$$= \pi^{-\frac{n}{2}}(1-\omega^2)^{-\frac{n}{2}} G_\omega(|x|^2 + |y|^2) \exp\left(\frac{2\omega}{1-\omega^2} x \cdot y\right).$$

As $\Phi_k(x,y)$ is the kernel of P_k it follows that

$$(3.4.6) \qquad \pi^{\frac{n}{2}}(1-\omega^2)^{\frac{n}{2}} \sum_{k=0}^\infty P_k f(x) \omega^k$$
$$= \int_{\mathbb{R}^n} G_\omega(|x|^2 + |y|^2) \exp\left(\frac{2\omega}{1-\omega^2} x \cdot y\right) f(y) dy.$$

Let us write $x = rx'$, $y = sy'$ where $r = |x|$, $s = |y|$ so that $P(y) = s^m P(y')$; then we have

$$(3.4.7) \qquad \pi^{n/2}(1-\omega^2)^{n/2} \sum_{k=0}^\infty P_k f(x) \omega^k$$
$$= G_\omega(r^2) \int_0^\infty \left(\int_{S^{n-1}} \exp\left(\frac{2\omega}{1-\omega^2} rsx' \cdot y'\right) P(y') dy' \right)$$
$$\times g(s) s^{n+m-1} ds$$

where $g(s) = G_\omega(s^2) f_0(s)$. To evaluate the integral we proceed as follows.

Using polar coordinates one has

$$\int_{\mathbb{R}^n} e^{-2\pi i x \cdot y} g(|y|) P(y) dy$$
$$= \int_0^\infty \left(\int_{S^{n-1}} e^{-2\pi i r s x' \cdot y'} P(y') dy' \right) g(s) s^{n+m-1} ds.$$

In view of the Hecke–Bochner identity for the Fourier transform

(3.4.8)
$$\int_0^\infty \left(\int_{S^{n-1}} e^{-2\pi i r s x' \cdot y'} P(y') dy' \right) g(s) s^{n+m-1} ds$$
$$= (2\pi) i^{-m} r^{-\frac{n}{2}+1} P(x')$$
$$\times \int_0^\infty g(s) J_{\frac{n}{2}+m-1}(2\pi r s) s^{\frac{n}{2}+m} ds.$$

Since both sides of this formula are holomorphic functions of r we can replace r by $\frac{i\omega r}{(1-\omega^2)\pi}$ which gives

(3.4.9)
$$\int_0^\infty \left(\int_{S^{n-1}} \exp\left(\frac{2\omega}{1-\omega^2} r s x' \cdot y'\right) P(y') dy' \right) g(s) s^{n+m-1} ds$$
$$= 2\pi^{\frac{n}{2}} (ir)^{-\frac{n}{2}-m+1} (1-\omega^2)^{-\frac{n}{2}+1} P(x)$$
$$\times \left(\int_0^\infty g(s) J_{\frac{n}{2}+m-1} \cdot \left(\frac{2i\omega}{1-\omega^2} r s\right) s^{\frac{n}{2}+m} ds \right).$$

Using this in (2.4.7) we obtain

(3.4.10)
$$\sum_{k=0}^\infty P_k f(x) \omega^k$$
$$= 2(1-\omega^2)^{-1} \omega^{-\frac{n}{2}+1} (ir)^{-\frac{n}{2}-m+1} P(x) G_\omega(r^2)$$
$$\cdot \left(\int_0^\infty g(s) J_{\frac{n}{2}+m-1}\left(\frac{2i\omega}{1-\omega^2} rs\right) s^{\frac{n}{2}+m} ds \right)$$

In view of the generating function identity (1.1.47) for $L_k^\delta(r^2) e^{-\frac{1}{2}r^2}$ namely,

$$\sum_{k=0}^\infty \frac{\Gamma(k+1)}{\Gamma(k+\delta+1)} L_k^\delta(r^2) L_k^\delta(s^2) e^{-\frac{1}{2}(r^2+s^2)} \omega^{2k}$$
$$= (1-\omega^2)^{-1} (irs\omega)^{-\delta} G_\omega(r^2+s^2) J_\delta\left(\frac{2i\omega}{1-\omega^2} rs\right),$$

we obtain the formula,

$$(3.4.11) \quad \sum_{k=0}^{\infty} P_k f(x) \omega^k$$

$$= P(x) \sum_{k=0}^{\infty} \frac{\Gamma(k+1)}{\Gamma(k+\delta+1)}$$

$$\times \left(\int_0^{\infty} f_0(s) L_k^{\delta}(s^2) e^{-\frac{1}{2}s^2} s^{n+2m-1} ds \right)$$

$$\times L_k^{\delta}(r^2) e^{-\frac{1}{2}r^2} \omega^{2k+m}$$

$$= P(x) \sum_{k=0}^{\infty} R_k^{\delta}(f_0) L_k^{\delta}(r^2) e^{-\frac{1}{2}r^2} \omega^{2k+m}$$

with $\delta = \frac{n}{2} + m - 1$. Comparing the coefficients on both sides we get the theorem. The corollary follows by taking $m = 0$. ∎

We can now establish the bounds (3.4.1) for the projection operators.

Proposition 3.4.1. *Assume that $f \in L^p(\mathbb{R}^n)$ is radial and $1 \leq p < \frac{2n}{n+1}$. Then the estimates (3.4.1) are valid.*

Proof: In view of Corollary 3.4.1 it follows that

$$(3.4.12) \quad \|P_{2k}f\|_2^2 = |R_k^{\frac{n}{2}-1}(f_0)|^2 \int_0^{\infty} |L_k^{\frac{n}{2}-1}(r^2)|^2 e^{-r^2} r^{n-1} dr.$$

Applying Holder's inequality to $R_k^{-\frac{n}{2}-1}(f_0)$ and using the estimate

$$(3.4.13) \quad \int_0^{\infty} |L_k^{\frac{n}{2}-1}(r^2)|^2 e^{-r^2} r^{n-1} dr \leq C k^{\frac{n}{2}-1}$$

we see that

$$(3.4.14) \quad \|P_{2k}f\|_2^2 \leq C k^{-\frac{n}{2}+1} \|f\|_p \left(\int_0^{\infty} |L_k^{\frac{n}{2}-1}(r^2) e^{-\frac{1}{2}r^2}|^{p'} r^{n-1} dr \right)^{1/p'}$$

where $\frac{1}{p} + \frac{1}{p'} = 1$. Now we can apply Lemma 1.5.4 to estimate the last integral which will then prove the proposition. ∎

Theorem 3.4.2. *Asume that $f \in L^p(\mathbb{R}^n)$ is radial and $\delta > 0$. Then*

$$\|S_R^{\delta} f\|_p \leq C \|f\|_p$$

holds for all p with $\frac{2n}{n+1+2\delta} < p < \frac{2n}{n-1-2\delta}$.

Proof: The proof follows from the above proposition and the estimate of Theorem 3.3.1. The argument is similar to the proof of Theorem 2.3.1 and the details are left to the interested reader. ∎

CHAPTER 4
MULTIPLIERS FOR HERMITE EXPANSIONS

In this chapter we prove a Marcinkiewicz type multiplier theorem for the Hermite expansions. Given a function m defined on the set of all nonnegative integers we look at the operator $m(H)$ defined via spectral theorem. We find conditions on the function m so that $m(H)$ is bounded on $L^p(\mathbb{R}^n)$, $1 < p < \infty$. In order to prove this result we develop and apply the Littlewood–Paley–Stein theory of g functions for the Hermite semigroup. We also study the mapping properties of the operators $A_j H^{-\frac{1}{2}}$ and $A_j^* H^{-\frac{1}{2}}$ which are the analogues of the classical Riesz transforms. Using the Riesz transform we study the solution operators to the wave equation associated to the Hermite operator.

4.1 Littlewood–Paley–Stein Theory for the Hermite Semigroup

In this section we study the g functions defined for the Hermite subgroup $T^t = e^{-tH}$. For $t > 0$, $T^t f$ is defined by the equation

(4.1.1) $$T^t f(x) = \sum_{k=0}^{\infty} e^{-(2k+n)t} P_k f(x).$$

In view of the Mehler's formula for the Hermite functions we can write (4.1.1) in the form

(4.1.2) $$T^t f(x) = (2\pi)^{-\frac{n}{2}} \int_{\mathbb{R}^n} K_t(x,y) f(y) dy$$

where the kernel $K_t(x,y)$ is explicitly given by

(4.1.3) $$K_t(x,y) = (\sinh 2t)^{-\frac{n}{2}} e^{-\frac{1}{2}(|x|^2+|y|^2)\coth 2t + x \cdot y \operatorname{cosech} 2t}.$$

Using this expression for the kernel it is not difficult to prove that T^t defines a strongly continuous semigroup which is self adjoint on $L^2(\mathbb{R}^n)$. We can also show that T^t is a contraction on $L^p(\mathbb{R}^n)$ for any $1 \leq p \leq \infty$.

As in Chapter 2 we now define the g_k functions by

(4.1.4) $$(g_k(f,x))^2 = \int_0^{\infty} |\partial_t^k T^t f(x)|^2 t^{2k-1} dt.$$

When $k = 1$ we write $g = g_k$. We remark that we have the relation

(4.1.5) $$g_k(f,x) \leq C_k g_{k+1}(f,x)$$

for any $k \geq 1$. This can be proved as in the case of the special Hermite semigroup. The first theorem we prove for the g_k functions is the following L^2 result.

Theorem 4.1.1. *For each $k \geq 1$ and $f \in L^2(\mathbb{R}^n)$ one has*

(4.1.6) $$\|g_k(f)\|_2^2 = 2^{-2k}\Gamma(2k)\|f\|_2^2.$$

Proof: From the definition it follows that

(4.1.7) $$\|g_k(f)\|_2^2 = \int_{\mathbb{R}^n} \int_0^\infty t^{2k-1} |\partial_t^k T^t f(x)|^2 dt\, dx.$$

Interchanging the order of integration and observing that

(4.1.8) $$\partial_t^k T^t f(x) = \sum_{j=0}^\infty (2j+n)^k (-1)^k e^{-(2j+n)t} P_j f(x)$$

we obtain, due to the orthogonality of the projections

(4.1.9) $$\int_{\mathbb{R}^n} |\partial_t^k T^t f(x)|^2 dx = \sum_{j=0}^\infty (2j+n)^{2k} e^{-(2j+n)2t} \|P_j f\|_2^2.$$

Integrating this against t^{2k-1} we complete the proof. ∎

We next proceed to prove that g_k are bounded on $L^p(\mathbb{R}^n)$, $1 < p < \infty$. In order to prove this it is enough to show that g_k are weak type $(1,1)$. For the sake of simplicity we consider the case $k = 1$. We will take $E_1 = \mathbb{C}$ and $E_2 = L^2(\mathbb{R}_+, t dt)$ and consider g as a singular integral operator taking values in E_2. If we let

(4.1.10) $$Tf(x) = \int_{\mathbb{R}^n} \partial_t K_t(x,y) f(y) dy$$

where K_t is as in (4.1.3) it follows that

(4.1.11) $$(g(f,x))^2 = (2\pi)^{-n} \|Tf\|_{E_2}^2,$$

the norm being that of E_2. Since g is bounded on $L^2(\mathbb{R}^n)$, we need to check that $\partial_t K_t(x,y)$ is a Calderon–Zygmund kernel taking values in the Hilbert space E_2. Then we can appeal to Theorem 2.3.2 to conclude that g is weak type $(1,1)$.

In order to show that $\partial_t K_t(x,y)$ is a Calderon–Zygmund kernel we need the following estimates.

Lemma 4.1.1.

(a) $\left|\frac{\partial}{\partial t}K_t(x,y)\right| \leq Ct^{-\frac{n}{2}-1}e^{-\frac{d}{t}|x-y|^2}$,

(b) $\left|\frac{\partial}{\partial y_j}\frac{\partial}{\partial t}K_t(x,y)\right| \leq Ct^{-\frac{n}{2}-\frac{3}{2}}e^{-\frac{d}{t}|x-y|^2}$, where $d > 0$ is a fixed constant.

This Lemma can be proved using the explicit formula (4.1.3) for the kernel $K_t(x,y)$. More estimates for the kernel K_t will be proved in the next section (see Lemma 4.3.1). As the proof of the above Lemma goes along similar lines we leave the proof to the interested reader. Here we assume the lemma and show how the estimates (a) and (b) imply that $\partial_t K_t(x,y)$ is a Calderon–Zygmund kernel. But that is easy. A simple calculation shows that (a) gives

$$(4.1.12) \qquad \int_0^\infty |\partial_t K_t(x,y)|^2 t\,dt \leq C|x-y|^{-2n}.$$

Similarly, the estimate (b) leads to

$$(4.1.13) \qquad \int_0^\infty \left|\frac{\partial}{\partial y_j}\frac{\partial}{\partial t}K_t(x,y)\right|^2 t\,dt \leq C|x-y|^{-2n-2}.$$

This proves that $\partial_t K_t$ is a Calderon–Zygmund kernel taking values in E_2.

We can now prove the following theorem regarding the L^p boundedness of the g function. For a later purpose we also need a weighted version. So, we recall the Muckenhoupt's A_p class of weight functions. For $1 < p < \infty$ we say that a nonnegative function $w(x)$ belongs to A_p if it satisfies

$$(4.1.14) \qquad \left(\frac{1}{|Q|}\int_Q w(y)dy\right)\left(\frac{1}{|Q|}\int_Q w(y)^{-1/p-1}dy\right)^{p-1} \leq C$$

for all cubes Q, where C is independent of Q. We denote by $\|f\|_{p,w}$ the weighted norm

$$(4.1.15) \qquad \|f\|_{p,w}^p = \int_{\mathbb{R}^n} |f(x)|^p w(x)dx.$$

Theorem 4.1.2. *There exist two constants C_1 and C_2 such that for all f in $L^p(\mathbb{R}^n)$, $1 < p < \infty$ the following inequalities are valid.*

(i) $C_1 \|f\|_p \leq \|g(f)\|_p \leq C_2 \|f\|_p$.

If $w(x) = |x|^{-n(\frac{p}{2}-1)}$ and $\frac{4}{3} < p < 4$ we also have the weighted inequalities

(ii) $C_1 \|f\|_{p,w} \leq \|g(f)\|_{p,w} \leq C_2 \|f\|_{p,w}$.

Proof: That g is bounded on L^p, $1 < p < \infty$ follows from the fact that the kernel $\partial_t K_t$ is a Calderon–Zygmund kernel. It is also known that singular integral operators satisfy weighted L^p inequality whenever $w \in A_p$. It can be checked that $|x|^{-n(\frac{p}{2}-1)}$ belongs to A_p for $\frac{4}{3} < p < 4$. Hence the direct inequality in (ii) also follows. The reverse inequalities are proved using Theorem 4.1.1. We prove $\|g(f)\|_{p,w} \geq C_1 \|f\|_{p,w}$. The unweighted case is simpler and so we omit the proof.

By polarizing the isometry $\|g(f)\|_2 = \frac{1}{2}\|f\|_2$ of Theorem 3.1.1 we obtain

$$(4.1.16) \quad \int f_1(x)f_2(x)dx = 4\int_{\mathbb{R}^n}\int_0^\infty t(\partial_t T f_1(x))(\partial_t T^t f_2(x))dtdx$$

which leads to the inequality

$$(4.1.17) \quad \left|\int_{\mathbb{R}^n} f_1(x)f_2(x)dx\right| \leq 4\int_{\mathbb{R}^n} g(f_1,x)g(f_2,x)dx.$$

Taking $h(x) = |x|^{-n(\frac{1}{2}-\frac{1}{p})}f_2(x)$ we get

$$\left|\int_{\mathbb{R}^n} f_1(x)|x|^{-n(\frac{1}{2}-\frac{1}{p})}f_2(x)dx\right|$$

$$\leq 4\int_{\mathbb{R}^n} g(f_2,x)|x|^{-n(\frac{1}{2}-\frac{1}{p})}|x|^{-n(\frac{1}{2}-\frac{1}{q})}g(h,x)dx$$

where $\frac{1}{p}+\frac{1}{q}=1$. By applying Holder's inequality we get

$$(4.1.18) \quad \left|\int f_1(x)|x|^{-n(\frac{1}{2}-\frac{1}{p})}f_2(x)dx\right| \leq C\|g(f_1)\|_{p,w}\|g(h)\|_{q,w}.$$

By the direct part of (ii) we have

$$(4.1.19) \quad \|g(h)\|_{q,w} \leq C\|h\|_{q,w} = C\|f_2\|_q.$$

Therefore, we have the inequality

$$(4.1.20) \quad \left|\int_{\mathbb{R}^n} f_1(x)|x|^{-n(\frac{1}{2}-\frac{1}{p})}f_2(x)dx\right| \leq C\|g(f_1)\|_{p,w}\|f_2\|_q.$$

Taking supremum over all f_2 with $\|f_2\| \leq 1$ we get

$$\|g(f_1)\|_{p,w} \geq C_1\|f_1\|_{p,w}. \qquad \blacksquare$$

In order to prove the multiplier theorem we need one more auxiliary function. For $k \geq 1$ we define

$$(4.1.21) \quad (g_k^*(f,x))^2$$

$$= \int_{\mathbb{R}^n}\int_0^\infty t^{-\frac{n}{2}+1}(1+t^{-1}|x-y|^2)^{-k}|\partial_t T^t f(y)|^2 dtdy.$$

Concerning the boundedness of g_k^* we prove the following result.

Theorem 4.1.3.
(i) If $p > 2$ and $k > \frac{n}{2}$ then
$$\|g_k^*(f)\|_p \leq C\|f\|_p.$$
(ii) If $2 < p < 4$ and $k > \frac{n}{2}$ then
$$\|g_k^*(f)\|_{p,w} \leq C\|f\|_{p,w}$$

where w is as in the previous theorem.

Proof: Since $k > \frac{n}{2}$ the function $(1+|x-y|^2)^{-k}$ is integrable and hence for $h \geq 0$

$$(4.1.22) \qquad \sup_{t>0} \int_{\mathbb{R}^n} t^{-n/2}(1+t^{-1}|x-y|^2)^{-k} h(y)dy \leq CMh(x)$$

where Mh is the Hardy–Littlewood maximal function. From this it follows that

$$(4.1.23) \qquad \int_{\mathbb{R}^n} (g_k^*(f,x))^2 h(x)dx \leq C \int_{\mathbb{R}^n} (g(f,x))^2 Mh(x)dx.$$

As M is bounded on $L^p(\mathbb{R}^n)$, $1 < p < \infty$ an application of Holder's inequality proves that

$$(4.1.24) \qquad \int_{\mathbb{R}^n} (g_k^*(f,x))^2 h(x)dx \leq C\|g(f)\|_p^2 \|h\|_s$$

where s is the conjugate exponent of $r = \frac{p}{2} > 1$. Taking supremum over all h the above proves (i).

To prove (ii) let us take $h_1(x) = h(x)|x|^{-n+\frac{n}{r}}$. (4.1.23) becomes

$$(4.1.25) \qquad \int_{\mathbb{R}^n} (g_k^*(f,x))^2 |x|^{-n+\frac{n}{r}} h(x)dx$$
$$\leq C \int_{\mathbb{R}^n} (g(f,x))^2 |x|^{-n+\frac{n}{r}} |x|^{\frac{n}{s}} Mh_1(x)dx.$$

Applying Holder's inequality

$$(4.1.26)$$
$$\int_{\mathbb{R}^n} (g(f,x))^2 |x|^{-n+\frac{n}{r}} |x|^{\frac{n}{s}} Mh_1(x)dx$$
$$\leq C\|g(f)\|_{p,w}^2 \int_{\mathbb{R}^n} |x|^n (Mh_1(x))^s dx.$$

As $p < 4$, $r < 2$ so that $s > 2$; this shows that $|x|^n \in A_s$. The Hardy-Littlewood maximal function satisfies

$$(4.1.27) \qquad \int (Mf(x))^p w(x) dx \leq C \int |f(x)|^p w(x) dx$$

whenever $w \in A_p$. Therefore, we get from (4.1.25) and (4.1.26)

(4.1.28)
$$\int_{\mathbb{R}^n} (g_k^*(f,x))^2 |x|^{-n+\frac{n}{r}} h(x) dx$$
$$\leq C \|f\|_{p,w}^2 \int_{\mathbb{R}^n} |x|^n |x|^{-ns+\frac{ns}{r}} (h(x))^s dx$$
$$= C\|f\|_{p,w}^2 \|h\|_s.$$

Again, taking supremum over all h we obtain (ii). This completes the proof. ∎

With these preparatory results we conclude this section. In the next section we take up the Marcinkiewicz multiplier theorem. ∎

4.2 Marcinkiewicz Multiplier Theorem

Given a function m defined and bounded on the set of all natural numbers we can use the spectral theorem to define $m(H)$. The action of $m(H)$ on a function f is given by

$$(4.2.1) \qquad m(H)f = \sum_{k=0}^{\infty} m(2k+n) P_k f.$$

This operator $m(H)$ is bounded on $L^2(\mathbb{R}^n)$. This follows immediately from the Plancherel theorem for the Hermite expansions as m is bounded. On the other hand, the mere boundedness of m is not sufficient to imply the L^p boundedness of $m(H)$ for $p \neq 2$. So, we need to impose further conditions on m to ensure that $m(H)$ is bounded on $L^p(\mathbb{R}^n)$.

More generally we can consider a function m defined and bounded on the set of all multiindices and define an operator T_m by the prescription

$$(4.2.2) \qquad T_m f = \sum_{\mu} m(\mu) \widehat{f}(\mu) \Phi_\mu.$$

We look for conditions on m. The sufficient conditions we find for m involve finite difference operators. As m is a function of n variables we define

(4.2.3) $$\Delta_j m(\mu) = m(\mu + e_j) - m(\mu).$$

Δ_j^k being defined inductively, we also set

(4.2.4) $$\Delta^\beta m(\mu) = \Delta_1^{\beta_1} \Delta_2^{\beta_2} \cdots \Delta_n^{\beta_n} m(\mu)$$

for any multiindex β. We are ready for

Theorem 4.2.1. *Assume that $k > \frac{n}{2}$ is an integer and the function m satisfies*

(4.2.5) $$|\Delta^\beta m(\mu)| \leq C_\beta (1 + |\mu|)^{-|\beta|}$$

for all β with $|\beta| \leq k$. Then the operator T_m is bounded on $L^p(\mathbb{R}^n)$, $1 < p < \infty$. Moreover, if $\frac{4}{3} < p < 4$ and $w(x) = |x|^{-n(p/2-1)}$ then we also have $\|T_m f\|_{p,w} \leq C \|f\|_{p,w}$.

Our conditions on the function m are similar to Hormander's condition for the Fourier multipliers. In proving this theorem we use the Littlewood–Paley–Stein theory developed in the previous section. We let $F(x) = T_m f(x)$ and show that when $k > \frac{n}{2}$,

(4.2.6) $$g_{k+1}(F, x) \leq C_k g_k^*(f, x)$$

for all x. Once we have this, the theorem for $p > 2$ follows from Theorems 4.1.2 and 4.1.3 in view of $g(F, x) \leq C g_{k+1}(F, x)$. By duality we obtain the theorem for $p < 2$. So, we only need to establish (4.2.6).

Using m we define the kernel

(4.2.7) $$M(t, x, y) = \sum_\mu e^{-(2|\mu|+n)t} m(\mu) \Phi_\mu(x) \Phi_\mu(y).$$

In order to prove (3.2.6) we need the following lemma which translates the conditions on m into properties of the kernel M.

Lemma 4.2.1. *Assume that m satisfies (4.2.5). Then we have*

(i) $|\partial_t^k M(t, x, y)| \leq C t^{-\frac{n}{2} - k}$

(ii) $\int_{\mathbb{R}^n} |x - y|^2 |\partial_t^k M(t, x, y)|^2 dy \leq C t^{-\frac{n}{2} - k}.$

Proof: The proof of (i) uses only the boundedness of m. In fact, applying Cauchy–Schwarz we see that

$$(4.2.8) \quad |\partial_t^k M(t,x,y)|^2 \leq C\Big(\sum_{N=0}^{\infty}(2N+n)^k e^{-(2N+n)t}\Phi_N(x,x)\Big)$$
$$\times \Big(\sum_{N=0}^{\infty}(2N+n)^k e^{-(2N+n)t}\Phi_N(y,y)\Big).$$

In view of Mehler's formula we have

$$(4.2.9) \quad \sum_{N=0}^{\infty}(2N+n)^k e^{-(2N+n)t}\Phi_N(x,x)$$
$$= (-1)^k \partial_t^k \big\{(2\pi)^{\frac{n}{2}}(\sinh 2t)^{-\frac{n}{2}} e^{-(\tanh t)|x|^2}\big\}$$

which is clearly bounded by $Ct^{-\frac{n}{2}-k}$. This proves (i).

In order to prove (ii) it is enough to show that

$$(4.2.10) \quad \int_{\mathbb{R}^n} |(x-y)^\beta \partial_t^k M(t,x,y)|^2 dy \leq Ct^{-\frac{n}{2}-k}$$

for any multiindex β with $|\beta| = k$. To prove this we write $\psi(\mu) = (-1)^k(2|\mu|+n)^k m(\mu) e^{-(2|\mu|+n)t}$ so that we have

$$(4.2.11) \quad \partial_t^k M(t,x,y) = M_\psi(x,y)$$

in the notation of Chapter 3. Now, we need the following formula:

$$(4.2.12) \quad (x-y)^\beta M_\psi(x,y) = \sum C_{\gamma\delta}(B-A)^\gamma \Delta^\delta M_\psi(x,y)$$

where the sum is extended over all the multiindices γ and δ satisfying $2\delta_j - \gamma_j = \beta_j$, $\delta_j \leq \beta_j$. This is just a variant of Lemma 3.2.3 and the proof is exactly the same except for some minor changes.

A typical term in the above sum (4.2.12) is of the form

$$(4.2.13) \quad \sum_\mu \Delta^\delta \psi(\mu) A^\tau \Phi_\mu(x) B^\sigma \Phi_\mu(y)$$

where $|\tau| + |\sigma| = |\gamma|$. The square of the L^2 norm of (4.2.13) is bounded by

$$(4.2.14) \quad \sum_\mu |\Delta^\delta \psi(\mu)|^2 (2|\mu|+n)^{|\gamma|}(\Phi_{\mu+|\tau|}(x))^2.$$

Now, applying Leibnitz rule for finite differences and making use of (4.2.5) it can be easily seen that

$$|\Delta^\delta \psi(\mu)| \leq C(2|\mu| + n)^{k-|\delta|} e^{-d(2|\mu|+n)t} \tag{4.2.15}$$

where $d > 0$ is some fixed constant. Therefore, (4.2.14) is bounded by

$$\sum (2|\mu| + n)^{2k-2|\delta|+|\gamma|} e^{-2d(2|\mu|+n)t} (\Phi_{\mu+|\tau|}(x))^2$$

which in turn is bounded by (as $2|\delta| - |\gamma| = |\beta| = k$)

$$\sum_{N=0}^{\infty} (2N + n)^k e^{-2d(2N+n)t} (\Phi_N(x))^2. \tag{4.2.16}$$

As before this gives the estimate $Ct^{-\frac{n}{2}-k}$.

Having proved the lemma we are now ready to establish (4.2.6). In terms of the kernel $M(t, x, y)$ we can write

$$T^{t+s} F(x) = \int_{\mathbb{R}^n} M(t, x, y) T^s f(y) dy \tag{4.2.17}$$

which can be verified easily. Differentiating (4.2.17) k times with respect to t, once with respect to s and then putting $t = s$ we get the relation

$$\partial_t^{k+1} T^{2t} F(x) = \int_{\mathbb{R}^n} \partial_t^k M(t, x, y) \partial_t T^t f(y) dy. \tag{4.2.18}$$

We write $\partial_t^k T^{2t} F(x) = A_t(x) + B_t(x)$ with

$$A_t(x) = \int_{|x-y| \leq t^{\frac{1}{2}}} \partial_t^k M(t, x, y) \partial_t T^t f(y) dy, \tag{4.2.19}$$

and

$$B_t(x) = \int_{|x-y| > t^{\frac{1}{2}}} \partial_t^k M(t, x, y) \partial_t T^t f(y) dy. \tag{4.2.20}$$

We will prove that $|A_t(x)|^2 + |B_t(x)|^2$ is bounded by a constant times

$$t^{-2k-\frac{n}{2}} \int_{\mathbb{R}^n} (1 + t^{-1}|x-y|^2)^{-k} |\partial_t T^t f(y)|^2 dy. \tag{4.2.21}$$

This will then prove that

$$(4.2.22) \quad \int_0^\infty |\partial_t^{k+1} T^t F(x)|^2 t^{2k+1} dx$$

$$\leq C \int_{\mathbb{R}^n} \int_0^\infty t^{-\frac{n}{2}+1}(1+t^{-1}|x-y|^2)^{-k} |\partial_t T^t f(y)|^2 dt dy$$

which establishes (4.2.6).

So, it remains to prove the estimate (4.2.21) for $|A_t(x)|^2$ and $|B_t(x)|^2$. First consider $A_t(x)$. Applying Cauchy–Schwarz and using the estimate (i) of Lemma 4.2.1 we have

$$(4.2.23) \quad |A_t(x)|^2 \leq C t^{-\frac{n}{2}-2k} \int_{|x-y|\leq t^{\frac{1}{2}}} |\partial_t T^t f(y)|^2 dy$$

$$\leq C t^{-\frac{n}{2}-2k} \int_{\mathbb{R}^n} (1+t^{-1}|x-y|^2)^{-k} |\partial_t T^t f(y)|^2 dy.$$

This proves the estimate for $A_t(x)$. Again $B_t(x)$ gives the estimate

$$(4.2.24) \quad |B_t(x)|^2 \leq \left(\int |x-y|^{2k} |\partial_t^k M(t,x,y)|^2 dy \right)$$

$$\times \left(\int_{|x-y|>t^{\frac{1}{2}}} |x-y|^{-2k} |\partial_t T^t f(y)|^2 dy \right).$$

Using the estimate (ii) of the lemma we get

$$(4.2.25) \quad |B_t(x)|^2 \leq C t^{-\frac{n}{2}-k} \int_{|x-y|>t^{\frac{1}{2}}} |x-y|^{-2k} |\partial_t T^t f(y)|^2 dy$$

$$\leq C t^{-\frac{n}{2}-2k} \int_{\mathbb{R}^n} (1+t^{-1}|x-y|^2)^{-k} |\partial_t T^t f(y)|^2 dy.$$

This is the estimate for $B_t(x)$. Hence (4.2.6) is established and this completes the proof of Theorem 4.2.1.

4.3 Conjugate Poisson Integrals and Riesz Transforms

Analogues of the classical conjugate harmonic functions and the conjugate mappings can be studied for a variety of classical orthogonal expansions. Here we propose to study them for the Hermite expansions. The classical Riesz transforms are the operators $\frac{\partial}{\partial x_j}(-\Delta)^{-\frac{1}{2}}$, $j = 1, 2, \ldots, n$. Since the Hermite operator can be expressed as

$$H = \tfrac{1}{2} \sum_{j=1}^n (A_j A_j^* + A_j^* A_j),$$

this suggests that we look at the operators $R_j = A_j H^{-\frac{1}{2}}$ and $R_j^* = A_j^* H^{-\frac{1}{2}}$ as the analogues of the Riesz transforms. For f in $L^2(\mathbb{R}^n)$, $R_j f$ and $R_j^* f$ have Hermite expansions

(4.3.1) $$R_j f = \sum_\mu (2\mu_j + 2)^{\frac{1}{2}} (2|\mu| + n)^{-\frac{1}{2}} \widehat{f}(\mu) \Phi_{\mu+e_j},$$

(4.3.2) $$R_j^* f = \sum_\mu (2\mu_j)^{\frac{1}{2}} (2|\mu| + n)^{-\frac{1}{2}} \widehat{f}(\mu) \Phi_{\mu-e_j}.$$

In the one dimensional case it is rather natural to look at the operators $A(H+1)^{-\frac{1}{2}}$ and $(H+1)^{-\frac{1}{2}} A^*$ as they turn out to be the shift operators:

(4.3.3) $$S_+ f = A(H+1)^{-\frac{1}{2}} f = \sum_{k=0}^\infty \widehat{f}(k) h_{k+1},$$

(4.3.4) $$S_- f = (H+1)^{-\frac{1}{2}} A^* f = \sum_{k=1}^\infty \widehat{f}(k) h_{k-1}.$$

The operators R_j, R_j^*, S_+ and S_- are all multiplier operators of special form which are clearly bounded on $L^2(\mathbb{R}^n)$. We are interested in the L^p boundedness of these operators.

With the intention of proving the L^p boundedness of R_j and R_j^* we introduce the conjugate Poisson integrals for the Hermite expansions. When $n \geq 2$ these are the operators defined by

(4.3.5) $$A_j H^{-\frac{1}{2}} e^{-tH^{\frac{1}{2}}} f$$
$$= \sum_\mu (2\mu_j + 2)^{\frac{1}{2}} (2|\mu| + n)^{-\frac{1}{2}} e^{-(2|\mu|+n)^{\frac{1}{2}} t} \widehat{f}(\mu) \Phi_{\mu+e_j},$$

(4.3.6) $$A_j^* H^{-\frac{1}{2}} e^{-tH^{\frac{1}{2}}} f$$
$$= \sum_\mu (2\mu_j)^{\frac{1}{2}} (2|\mu| + n)^{-\frac{1}{2}} e^{-(2|\mu|+n)^{\frac{1}{2}} t} \widehat{f}(\mu) \Phi_{\mu-e_j}.$$

When $n = 1$ we consider the conjugate Poisson integrals defined in the way

(4.3.7) $$S_+^t f = A(H+1)^{-\frac{1}{2}} e^{-t(H+1)^{\frac{1}{2}}} f,$$

(4.3.8) $$S_-^t f = (H+1)^{-\frac{1}{2}} A^* e^{-t(H+1)^{\frac{1}{2}}} f.$$

We also consider the maximal conjugate Poisson integrals. When $n = 1$ they are defined by

$$(4.3.9) \qquad S_+^* f(x) = \sup_{0<t<1} |S_+^t f(x)|,$$

$$(4.3.10) \qquad S_-^* f(x) = \sup_{0<t<1} |S_-^t f(x)|,$$

When $n \geq 2$ the maximal conjugate Poisson integrals are defined by

$$(4.3.11) \qquad U_j f(x) = \sup_{0<t<1} |A_j H^{-\frac{1}{2}} e^{-tH^{\frac{1}{2}}} f(x)|,$$

$$(4.3.12) \qquad U_j^* f(x) = \sup_{0<t<1} |A_j^* H^{-\frac{1}{2}} e^{-tH^{\frac{1}{2}}} f(x)|.$$

We prove that S_+^* and S_-^* are bounded on $L^2(\mathbb{R})$ and weak type $(1,1)$ and when $n \geq 2$, both U_j and U_j^* are bounded on $L^2(\mathbb{R}^n)$ and weak type $(1,1)$. The weak type inequality will enable us to define $R_j f$ and $R_j^* f$ as the limits of $A_j H^{-\frac{1}{2}} e^{-tH^{\frac{1}{2}}} f$ and $A_j^* H^{-\frac{1}{2}} e^{-tH^{\frac{1}{2}}} f$ as $t \to 0$ for f in $L^1(\mathbb{R}^n)$. Likewise, we can define $S_+ f$ and $S_- f$ for functions f in $L^1(\mathbb{R})$.

We first consider the L^2 boundedness of the maximal conjugate Poisson integrals. The kernel of the Hermite semigroup e^{-tH} is $(2\pi)^{-\frac{n}{2}} K_t(x, y)$ where $K_t(x, y)$ is given by (4.1.3). By letting

$$(4.3.13) \qquad \varphi(t, x, y) = \tfrac{1}{2}|x-y|^2 \coth 2t + x \cdot y \tanh t$$

the kernel can be written as

$$(4.3.14) \qquad K_t(x, y) = (\sinh 2t)^{-\frac{n}{2}} e^{-\varphi(t,x,y)}.$$

Also, the function $\tfrac{1}{4}|x-y|^2 \coth 2t + x \cdot y \tanh t$ can be easily seen to be positive and hence

$$(4.3.15) \qquad K_t(x, y) \leq (\sinh 2t)^{-\frac{n}{2}} e^{-\frac{1}{4}|x-y|^2 \coth 2t}.$$

We make use of this estimate in the following proposition regarding the L^2 boundedness of the maximal conjugate Poisson integrals.

Proposition 4.3.1. *The maximal conjugate Poisson integrals are all bounded on $L^2(\mathbb{R}^n)$.*

Proof: We first consider the case $n \geq 2$. From the definition it follows that

(4.3.16) $$e^{-t(H-2)^{\frac{1}{2}}}(A_j H^{-\frac{1}{2}} f) = A_j H^{-\frac{1}{2}}(e^{-tH^{\frac{1}{2}}} f).$$

This gives us

(4.3.17) $$U_j f(x) = \sup_{0<t<1} |e^{-t(H-2)^{\frac{1}{2}}}(A_j H^{-\frac{1}{2}} f(x))|.$$

Since $A_j H^{-\frac{1}{2}}$ is bounded on $L^2(\mathbb{R}^n)$, it is enough to show that

(4.3.18) $$\sup_{0<t<1} |e^{-t(H-2)^{\frac{1}{2}}} f(x)| \leq CMf(x)$$

where Mf is the Hardy–Littlewood maximal function of f. In view of the subordinate identity

(4.3.19) $$e^{-\alpha} = \pi^{-\frac{1}{2}} \int_0^\infty e^{-\frac{1}{u}} u^{-\frac{3}{2}} e^{-\frac{u}{4}\alpha^2} du$$

we have the formula

(4.3.20) $$e^{-t(H-2)^{\frac{1}{2}}} = \pi^{-\frac{1}{2}} \int_0^\infty e^{-\frac{1}{u}} u^{-\frac{3}{2}} e^{-\frac{ut^2}{4}(H-2)} du.$$

As we are assuming $n \geq 2$, (4.3.18) would follow once we show that

(4.3.21) $$\sup_{0<u<\infty} |e^{-u(H-n)} f(x)| \leq CMf(x).$$

The kernel of $e^{-u(H-n)}$ is $(2\pi)^{-\frac{n}{2}} e^{un} K_u(x,y)$ which is bounded by a constant times

$$(1 + e^{-4u})^{-\frac{n}{2}} (\coth 2u)^{-\frac{n}{2}} e^{-\frac{1}{4}|x-y|^2 \coth 2u}.$$

From this estimate (4.3.21) follows immediately.

The proof that S_+^* and S_-^* are bounded on $L^2(\mathbb{R})$ is similar to the above. We make use of the identities

(4.3.22) $$S_+^t f(x) = e^{-tH^{\frac{1}{2}}}(S_+ f(x)),$$

(4.3.23) $$S_-^t f(x) = e^{-t(H+3)^{\frac{1}{2}}}(S_- f(x)).$$

We can proceed as above to show that $S_+^* f$ and $S_-^* f$ are dominated by $M(S_+ f)$ and $M(S_- f)$ respectively. ∎

To prove that the maximal conjugate Poisson integrals are weak type $(1,1)$ we need several estimates for the kernel K_t and its derivatives.

Lemma 4.3.1. *There exist positive constants C and a independent of x, y and t, $0 < t < 1$ such that the following estimates are valid.*

(i) $|\frac{\partial}{\partial x_j} K_t(x,y)| \leq Ct^{-\frac{(n+1)}{2}} e^{-\frac{a}{t}|x-y|^2}$,

(ii) $|x_j K_t(x,y)| \leq Ct^{-\frac{(n+1)}{2}} e^{-\frac{a}{t}|x-y|^2}$,

(iii) $|x_j \frac{\partial}{\partial y_i} K_t(x,y)| \leq Ct^{-\frac{n}{2}-1} e^{-\frac{a}{t}|x-y|^2}$,

(iv) $|\frac{\partial^2}{\partial x_j \partial y_i} K_t(x,y)| \leq Ct^{-\frac{n}{2}-1} e^{-\frac{a}{t}|x-y|^2}$.

Proof: Since the kernel $K_t(x,y)$ is the product of the one dimensional kernels $K_t(x_j, y_j)$ it is enough to consider the lemma when $n = 1$. First consider

$$(4.3.24) \qquad \frac{\partial}{\partial x} K_t(x,y) = (\sinh 2t)^{-\frac{1}{2}} e^{-\varphi(t,x,y)} (y \operatorname{cosech} 2t - x \coth 2t)$$

which can be written as the sum of

$$(4.3.25) \qquad A_t(x,y) = (\sinh 2t)^{-\frac{3}{2}} (y - x) e^{-\varphi(t,x,y)},$$

and

$$(4.3.26) \qquad B_t(x,y) = -2(\sinh t)^2 (\sinh 2t)^{-\frac{3}{2}} x e^{-\varphi(t,x,y)}.$$

For $0 < t < 1$, $\sinh t$ behaves like t and $\coth t$ behaves like t^{-1}. Therefore, using (4.3.15) we see that

$$(4.3.27) \qquad |A_t(x,y)| \leq Ct^{-\frac{3}{2}} |x - y| e^{-\frac{1}{8t}|x-y|^2} \leq Ct^{-1} e^{-\frac{a}{t}|x-y|^2}$$

with $a = \frac{1}{16}$. To estimate $B_t(x,y)$ it is enough to consider the kernel $xK_t(x,y)$ and prove (ii).

To estimate $xK_t(x,y)$ we first assume that $|x| \leq 4|y|$. Then we can write

$$(4.3.28) \qquad |xK_t(x,y)| \leq Ct^{-\frac{1}{2}} |2xy|^{\frac{1}{2}} e^{-\varphi(t,x,y)}.$$

When $xy \geq 0$, $(xy)^{\frac{1}{2}} e^{-xy \tanh t}$ is bounded by a constant times $t^{-\frac{1}{2}}$ and hence

$$(4.3.29) \qquad |xK_t(x,y)| \leq Ct^{-1} e^{-\frac{1}{4t}|x-y|^2}.$$

On the other hand, when $xy < 0$, $|2xy| = -2xy \leq (x-y)^2$ and so

$$(4.3.30) \qquad |xK_t(x,y)| \leq Ct^{-\frac{1}{2}} |x - y| e^{-\frac{1}{8t}|x-y|^2} \leq Ct^{-1} e^{-\frac{1}{16t}|x-y|^2}.$$

This proves the estimate (ii) when $|x| \leq 4|y|$. Next assume that $|x| > 4|y|$. We claim that

(4.3.31) $\quad\quad\quad \frac{1}{2}\varphi(t,x,y) \geq \frac{1}{8}x^2(\coth 2t).$

When $xy \geq 0$ this follows from

(4.3.32) $\quad\quad \frac{1}{2}\varphi(t,x,y) = \frac{1}{4}(x^2+y^2)\coth 2t - \frac{1}{2}xy\,\text{cosech}\,2t$

$$\geq \tfrac{1}{4}(\coth 2t)(x^2 - 2xy) \geq \tfrac{1}{8}x^2(\coth 2t).$$

When $xy < 0$ it is clear that $\frac{1}{2}\varphi(t,x,y) \geq \frac{1}{4}x^2(\coth 2t)$. Therefore, we have

$$|xK_t(x,y)| \leq Ct^{-\frac{1}{2}}|x|e^{-\frac{1}{2}\varphi(t,x,y)}e^{-\frac{1}{2}\varphi(t,x,y)} \leq Ct^{-1}e^{-\frac{1}{8t}|x-y|^2}.$$

The estimates (iii) and (iv) are proved in a similar fashion. We leave the details to the interested reader. ∎

Lemma 4.3.2. *For $t \geq 1$ the following estimates are valid with two positive constants C and b independent of x, y and t.*

(i) $\quad |x_j K_t(x,y)| \leq Ce^{-nt}e^{-b|x-y|^2}$
(ii) $\quad |\frac{\partial}{\partial x_j}K_t(x,y)| \leq Ce^{-nt}e^{-b|x-y|^2}.$

The proof of this lemma is very similar to that of the previous lemma. We have to use the fact that for $t \geq 1$ both $\sinh 2t$ and $\cosh 2t$ behave like e^{2t}. The details are omitted. ∎

Theorem 4.3.1. *The maximal conjugate Poisson integrals are weak type $(1,1)$. Consequently, they are bounded on $L^p(\mathbb{R}^n)$, $1 < p < \infty$.*

Proof: We consider the case $n \geq 2$. The case $n = 1$ is similar and so will be omitted. We imitate the standard proof of the weak type inequality for Calderon-Zygmund singular integrals. Given f in $L^1(\mathbb{R}^n)$ we take the Calderon-Zygmund decomposition $f = g + b$. Suppose we are given a singular integral operator T defined by a kernel $K(x,y)$,

(4.3.33) $\quad\quad\quad Tf(x) = \int_{\mathbb{R}^n} K(x,y)f(y)dy.$

If we assume that T is bounded on $L^2(\mathbb{R}^n)$ then the term Tg is taken care of; i.e.,

(4.3.34) $\quad\quad\quad |\{x : |Tg(x)| > \lambda\}| \leq C\lambda^{-1}\|f\|_1$

is true. To obtain the same inequality for Tb what we really need is the estimate

(4.3.35) $$\left|\frac{\partial}{\partial y_j}K(x,y)\right| \leq C|x-y|^{-n-1}$$

for the derivatives of the kernel $K(x,y)$. Then the proof of the weak type inequality for Tb is well known.

Suppose now we have a kernel $k_t(x,y)$ depending on a parameter t and suppose we are interested in the weak type inequality for the maximal operator

$$\sup_{0<t<1}|T_tf(x)| = \sup_{0<t<1}\left|\int k_t(x,y)f(y)dy\right|.$$

If we know that $\sup_{0<t<1}|T_tf|$ is bounded on $L^2(\mathbb{R}^n)$ then as before the term $\sup_{0<t<1}|T_tg|$ is taken care of. The weak type inequality for $\sup_{0<t<1}|T_tb|$ can be established if we know that

$$\sup_{0<t<1}\left|\frac{\partial}{\partial y_j}k_t(x,y)\right| \leq C|x-y|^{-n-1}$$

with C independent of t. The proof is simply an imitation of the proof of the t-independent case.

Now the conjugate Poisson integral $A_j H^{-\frac{1}{2}}e^{-tH^{\frac{1}{2}}}$ is given by a kernel $K_t^j(x,y)$. This kernel can be calculated in the following way. Differentiating the subordinate identity (4.3.19) with respect to α we obtain

(4.3.36) $$\alpha^{-1}e^{-\alpha} = \frac{1}{2\sqrt{\pi}}\int_0^\infty e^{-\frac{1}{u}}u^{-\frac{1}{2}}e^{-\frac{u}{4}\alpha^2}du.$$

From this we obtain the formula

(4.3.37) $$H^{-\frac{1}{2}}e^{-tH^{\frac{1}{2}}} = \frac{1}{\sqrt{\pi}}\int_0^\infty e^{-\frac{t^2}{4u}}u^{-\frac{1}{2}}e^{-uH}du.$$

Thus the kernel K_t^j is given by

(4.3.38) $$K_t^j(x,y) = \sqrt{2}(2\pi)^{-\frac{(n+1)}{2}}\int_0^\infty e^{-\frac{t^2}{4u}}u^{-\frac{1}{2}}\left(-\frac{\partial}{\partial x_j}+x_j\right)K_u(x,y)du.$$

In view of Proposition 4.3.1 we need to consider only the term b in the Calderon–Zygmund decomposition. Let us write

(4.3.39) $$\int K_t^j(x,y)b(y)dy = \int L_t^j(x,y)b(y)dy + \int M_t^j(x,y)b(y)dy,$$

where L_t^j and M_t^j are the kernels

(4.3.40) $\quad L_t^j(x,y) = \sqrt{2}(2\pi)^{-\frac{(n+1)}{2}} \int_0^1 e^{-\frac{t^2}{4u}} u^{-\frac{1}{2}} \left(-\frac{\partial}{\partial x_j} + x_j\right) K_u(x,y) du,$

(4.3.41) $\quad M_t^j(x,y) = \sqrt{2}(2\pi)^{-\frac{(n+1)}{2}} \int_1^\infty e^{-\frac{t^2}{4u}} u^{-\frac{1}{2}} \left(-\frac{\partial}{\partial x_j} + x_j\right) K_u(x,y) du.$

From the estimates of Lemma 4.3.2 it follows that

(4.3.42) $\quad \sup_{0<t<1} \left| \int M_t^j(x,y) b(y) dy \right| \leq C \int e^{-b|x-y|^2} |b(y)| dy$

and so it is immediate that this term satisfies the weak type inequality. From the estimates of Lemma 4.3.1 it follows that

(4.3.43) $\quad \sup_{0<t<1} \left|\frac{\partial}{\partial y_j} L_t^j(x,y)\right| \leq C \int_0^1 u^{-\frac{1}{2}} u^{-\frac{n}{2}-1} e^{-\frac{a}{u}|x-y|^2} du$

$\leq C|x-y|^{-n-1}$

with C independent of t. Therefore, by the foregoing remarks it is clear that the term corresponding to L_t^j is also of weak type $(1,1)$. This completes the proof of the weak type inequality for U_j. By the Marcinkiewicz interpolation theorem it follows that they are bounded on $L^p(\mathbb{R}^n)$, $1 < p < \infty$. ∎

We now state the main theorem concerning the Riesz transforms.

Theorem 4.3.2. *For $1 < p < \infty$ the Riesz transforms R_j and R_j^* are bounded on $L^p(\mathbb{R}^n)$. When $n \geq 2$ they are also weak type $(1,1)$. When $n = 1$ the shift operators S_+ and S_- are weak type $(1,1)$.*

Proof: The weak type inequality for the maximal conjugate Poisson integrals allow us to define $R_j f$ and $R_j^* f$ ($S_+ f$ and $S_- f$ when $n=1$) for functions in $L^1(\mathbb{R}^n)$. In fact, when f is a finite linear combination of Φ_μ, the conjugate Poisson integral $A_j H^{-\frac{1}{2}} e^{-tH^{\frac{1}{2}}} f$ converges to $A_j H^{-\frac{1}{2}} f$ as $t \to 0$. Then it follows that the limit

(4.3.44) $\quad R_j f(x) = \lim_{t \to 0} A_j H^{-\frac{1}{2}} e^{-tH^{\frac{1}{2}}} f(x)$

exists for every f in $L^1(\mathbb{R}^n)$. We take this to be the definition of R_j. Clearly, R_j is then of weak type $(1,1)$ and hence bounded on $L^p(\mathbb{R}^n)$, for $1 < p < \infty$ when $n \geq 2$. Similarly, $S_+ f$ and $S_- f$ are defined for

$f \in L^1(\mathbb{R})$ and are weak type $(1,1)$. Since the operator $(H+1)^{\frac{1}{2}} H^{-\frac{1}{2}}$ is bounded on $L^p(\mathbb{R})$, $1 < p < \infty$ by the Marcinkiewicz multiplier theorem it follows that the operators $AH^{-\frac{1}{2}}$ and $A^* H^{-\frac{1}{2}}$ are bounded on $L^p(\mathbb{R})$, $1 < p < \infty$. ∎

In the case of classical Riesz transforms it is well known that they are bounded on the Hardy space $H^1(\mathbb{R}^n)$. In fact, $H^1(\mathbb{R}^n)$ is characterized by the Riesz transforms:

(4.3.45) $$H^1(\mathbb{R}^n) = \Big\{ f \in L^1(\mathbb{R}^n) : \frac{\partial}{\partial x_j}(-\Delta)^{-\frac{1}{2}} f \in L^1(\mathbb{R}^n), \\ j = 1, 2, \ldots, n \Big\}.$$

In the present case we show that our Riesz transforms R_j and R_j^* are bounded on the local Hardy space $h^1(\mathbb{R}^n)$. It would be interesting to see if $h^1(\mathbb{R}^n)$ is characterized by R_j, $j = 1, 2, \ldots, n$, i.e., we like to know if the equality

(4.3.46) $$h^1(\mathbb{R}^n) = \{ f \in L^1(\mathbb{R}^n) : R_j f \in L^1(\mathbb{R}^n), j = 1, 2, \ldots, n \}$$

is true or not.

We briefly recall the definition and some properties of $h^1(\mathbb{R}^n)$. These spaces were introduced and studied by Goldberg [Go]. We take a nonnegative function φ in $\mathcal{S}(\mathbb{R}^n)$, and define $\varphi_t(x) = t^{-n} \varphi(\frac{x}{t})$, $t > 0$. Then

(4.3.47) $$h^1(\mathbb{R}^n) = \Big\{ f \in L^1(\mathbb{R}^n) : \sup_{0 < t < 1} |f * \varphi_t| \in L^1(\mathbb{R}^n) \Big\}.$$

Consider the symbol class $S_{1,0}^m$ which is defined to be the set of all $\sigma \in C^\infty(\mathbb{R}^n \times \mathbb{R}^n)$ which satisfy the estimates

(4.3.48) $$|\partial_x^\alpha \partial_\xi^\beta \sigma(x, \xi)| \leq C(1 + |\xi|)^{m - |\alpha| - |\beta|},$$

for all α and β with a constant C independent of x. Our definition differs from the usual one in that we are taking C to be independent of x. For σ in $S_{1,0}^m$ the pseudo differential operator $\sigma(x, D)$ is defined by

(4.3.49) $$\sigma(x, D) f(x) = (2\pi)^{-\frac{n}{2}} \int e^{ix \cdot \xi} \sigma(x, \xi) \hat{f}(\xi) d\xi.$$

It is known that when $\sigma \in S_{1,0}^0$ then $\sigma(x, D)$ is bounded on $h^1(\mathbb{R}^n)$. We now state and prove the following result.

Theorem 4.3.3. *The Riesz transforms R_j and R_j^* are all bounded on $h^1(\mathbb{R}^n)$.*

Proof: We merely verify that R_j and R_j^* are pseudodifferential operators whose symbols are in the class $S_{1,0}^0$. Since the symbol of A_j is $(-\xi_j + x_j)$ it is enough to show that the symbol $a(x,\xi)$ of $H^{-\frac{1}{2}}$ satisfies

$$(4.3.50) \qquad |\partial_\xi^\alpha \partial_x^\beta a(x,\xi)| \leq C(1+|x|+|\beta|)^{-1-|\alpha|-|\beta|}.$$

To this end we write

$$(4.3.51) \qquad H^{-\frac{1}{2}} = \frac{1}{\sqrt{\pi}} \int_0^\infty t^{-\frac{1}{2}} e^{-tH} dt$$

so that the symbol $a(x,\xi)$ is given by

$$(4.3.52) \qquad a(x,\xi) = \frac{1}{\sqrt{\pi}} \int_0^\infty t^{-\frac{1}{2}} \sigma_t(x,\xi) dt$$

where $\sigma_t(x,\xi)$ is the symbol of e^{-tH}. Now we can calculate $\sigma_t(x,\xi)$ as follows.

For f in the Schwartz class $\mathcal{S}(\mathbb{R}^n)$, $e^{-tH}f$ is given by

$$(4.3.53) \qquad e^{-tH} f = \sum_\mu e^{-(2|\mu|+n)t} (f, \Phi_\mu) \Phi_\mu.$$

The relation $\widehat{\Phi}_\mu = (-i)^{|\mu|} \Phi_\mu$ shows that

$$(4.3.54) \qquad (f, \Phi_\mu) = (\widehat{f}, \widehat{\Phi}_\mu) = (i)^{|\mu|} (\widehat{f}, \Phi_\mu).$$

Therefore, we can write

$$(4.3.55) \qquad e^{-tH} f = e^{in\pi/4} \sum_\mu e^{-(2|\mu|+n)(t+i\frac{\pi}{4})} (\widehat{f}, \Phi_\mu) \Phi_\mu.$$

From this it follows that

$$(4.3.56) \qquad \sigma_t(x,\xi) = e^{in\pi/4} e^{-ix\cdot\xi} \sum_\mu e^{-(2|\mu|+n)(t+i\frac{\pi}{4})} \Phi_\mu(x) \Phi_\mu(\xi).$$

Using Mehler's formula, we get

$$(4.3.57) \qquad \sigma_t(x,\xi) = e^{in\pi/4} (2\pi)^{-n/2} (\cosh 2t)^{-\frac{n}{2}} e^{-b(t,x,\xi)}$$

where

(4.3.58) $\quad b(t,x,\xi) = \frac{1}{2}(|x|^2 + |\xi|^2)\tanh 2t + 2ix \cdot \xi(\operatorname{sech} 2t)(\sinh t)^2$.

Thus the symbol $a(x,\xi)$ is given by

(4.3.59) $\quad a(x,\xi) = c_0 \int_0^\infty t^{-\frac{1}{2}}(\cosh 2t)^{-\frac{n}{2}} e^{-b(t,x,\xi)} dt$.

From the form of $b(t,x,\xi)$ it is clear that

(4.3.60) $\quad a_2(x,\xi) = c_0 \int_1^\infty t^{-\frac{1}{2}}(\cosh 2t)^{-\frac{n}{2}} e^{-b(t,x,\xi)} dt$,

is a rapidly decreasing function of $(|x| + |\xi|)$ and so are its derivatives. So, in order to prove that $a(x,\xi)$ satisfies (4.3.50) it is enough to prove the following lemma. ■

Lemma 4.3.4. *Let $a_1(x,\xi)$ be defined by*

(4.3.61) $\quad a_1(x,\xi) = c_0 \int_0^1 t^{-\frac{1}{2}}(\cosh 2t)^{-\frac{n}{2}} e^{-b(t,x,\xi)} dt$.

Then for any multiindices α and β we have

(4.3.62) $\quad |\partial_x^\alpha \partial_\xi^\beta a_1(x,\xi)| \leq C(1 + |\xi| + |x|)^{-1-|\alpha|-|\beta|}$.

Proof: Since $0 < t < 1$, $\tanh 2t$ behaves like t and $\cosh 2t = 0(1)$. Therefore,

(4.3.63) $\quad |a_1(x,\xi)| \leq C \int_0^1 t^{-\frac{1}{2}} e^{-\frac{1}{2}(|x|^2+|\xi|^2)t} dt \leq C(1 + |x| + |\xi|)^{-1}$.

When we take a derivative, say $\frac{\partial}{\partial \xi_j} a_1(x,\xi)$, we have two terms. The term

(4.3.64) $\quad \xi_j \int_0^1 t^{-\frac{1}{2}}(\cosh 2t)^{-\frac{n}{2}}(\tanh 2t) e^{-b(b,x,\xi)} dt$

is bounded by

$$|\xi|(1 + |x|^2 + |\xi|^2)^{-\frac{3}{2}} \leq C(1 + |x| + |\xi|)^{-2}.$$

The other term has a better decay as the derivative falling on $x \cdot \xi(\sinh t)^2(\operatorname{sech} 2t)$ brings down $x_j(\sinh t)^2(\operatorname{sech} 2t)$. Derivatives with respect to x and higher order derivatives are estimated in a similar fashion. This completes the proof of the lemma. ■

4.4 Wave Equation for the Hermite Operator

In this section we study the solution operators for the wave equation associated to the Hermite operator $H = (-\Delta + |x|^2)$ on $\mathbb{R}^n$. We first consider the following Cauchy problem:

(4.4.1) $$\partial_t^2 u(x,t) = (-\Delta + |x|^2)u(x,t),$$

(4.4.2) $$u(x,0) = 0, \quad \partial_t u(x,0) = f(x).$$

Formally, the unique solution of the above problem is given by

(4.4.3) $$u(x,t) = H^{-\frac{1}{2}} \sin(tH^{\frac{1}{2}})f(x).$$

The solution u has the Hermite expansion given by

(4.4.4) $$u(x,t) = \sum_{k=0}^{\infty} (2k+n)^{-\frac{1}{2}} \sin(2k+n)^{\frac{1}{2}} t P_k f(x).$$

In general, the above series defining u need not converge when $f \in L^p(\mathbb{R}^n)$, $p \neq 2$. Nevertheless, we prove below that the series converges in the L^2 norm whenever $f \in L^p(\mathbb{R}^n)$, provided p is close enough to 2. Thus for initial L^p data we get the solution in L^2 and the dependence of u on f is continuous. More precisely we have the following result.

Theorem 4.4.1. *Assume that $f \in L^p(\mathbb{R}^n)$, $\frac{2(n+1)}{n+3} < p \leq 2$. Then the solution u defined in (4.4.3) satisfies*

$$\|u(\cdot,t)\|_2 \leq C\|f\|_p$$

where C is a constant independent of t.

Proof: We prove the theorem using analytic interpolation. Recall that

$$J_{1/2}(s) = \left(\frac{2}{\pi}\right)^{\frac{1}{2}} s^{-\frac{1}{2}} \sin s.$$

In view of this we consider the family

(4.4.5) $$S_t(\alpha) = \left(\frac{\pi}{2}\right)^{\frac{1}{2}} t(tH^{\frac{1}{2}})^{\alpha - \frac{n}{2}} J_{\frac{n}{2} - \alpha}(tH^{\frac{1}{2}})$$

where α can be taken to be complex. Thus we have an analytic family of operators and it can be checked that this family is admissible. Moreover, when $\alpha = \frac{n-1}{2}$ we have

(4.4.6) $$S_t\left(\frac{n-1}{2}\right) = H^{-\frac{1}{2}} \sin(tH^{\frac{1}{2}}).$$

For the analytic family $S_t(\alpha)$ we first prove the following proposition.

Proposition 4.4.1.

(i) $\left\|S_t\left(\frac{n+1}{2} + iy\right)f\right\|_2 \leq C(y)\|f\|_2$

(ii) $\|S_t(-s+iy)f\|_2 \leq C(y)t^{-s-(\frac{n-1}{2})}\|f\|_1$

for every $s > 0$. Here $C(y)$ is of admissible growth.

Assuming the proposition for a moment we complete the proof of the theorem. We define

(4.4.7) $\qquad T_t(\alpha) = S_t\left(-s + \left(\frac{n+1}{2} + s\right)\alpha\right), \ s > 0.$

Then from the proposition it follows that

(4.4.8) $\qquad \|T_t(iy)\|_2 \leq C(y)t^{-s-(\frac{n-1}{2})}\|f\|_1,$

(4.4.9) $\qquad \|T_t(1+iy)f\|_2 \leq C(y)t\|f\|_2.$

By analytic interpolation theorem we conclude that for $0 < \alpha < 1$

(4.4.10) $\qquad \|T_t(\alpha)f\|_2 \leq C\|f\|_p$

where p is defined by $\frac{1}{p} = 1 - \frac{\alpha}{2}$. By choosing $\alpha = \frac{n-1+2s}{n+1+2s}$ we get $T_k(\alpha) = S_t\left(\frac{n-1}{2}\right)$ and hence

(4.4.11) $\qquad \|H^{-\frac{1}{2}}\sin(tH^{\frac{1}{2}})f\|_2 \leq C\|f\|_{p_s}$

where $\frac{1}{p_s} = \frac{n+3+2s}{2(n+1+2s)}$. If $p > \frac{2(n+1)}{n+3}$ is given we can choose $s > 0$ so that $p > p_s$ and $H^{-\frac{1}{2}}\sin tH^{\frac{1}{2}}$ is bounded from L^{p_s} into L^2. Then by the Riesz–Thorin convexity theorem we get

(4.4.12) $\qquad \|H^{-\frac{1}{2}}\sin(tH^{\frac{1}{2}})f\|_2 \leq C\|f\|_p.$

This proves the theorem.

Coming back to the proposition we see that (i) is immediate. In fact

(4.4.13) $\quad S_t\left(\frac{n+1}{2}\right)f = \left(\frac{\pi}{2}\right)^{\frac{1}{2}} t \sum_{k=0}^{\infty}((2k+n)t^2)^{\frac{1}{4}} J_{-\frac{1}{2}}((2k+n)^{\frac{1}{2}}t) P_k f.$

As $|J_\alpha(t)| \leq Ct^{-\frac{1}{2}}$ for $t \geq 1$ we see that $(2k+n)^{\frac{1}{4}}t^{\frac{1}{2}}|J_{-\frac{1}{2}}((2k+n)^{\frac{1}{2}}t)| \leq C$ for all k. Hence the operator $S_t\left(\frac{n+1}{2}\right)$ is bounded on $L^2(\mathbb{R}^n)$. To prove (ii) of the proposition we need the following estimate for the projections P_k.

Proposition 4.4.2. For $f \in L^p(\mathbb{R}^n)$, $1 \leq p \leq 2$,

$$\|P_k f\|_2 \leq C k^{\frac{n-1}{2}(\frac{1}{p} - \frac{1}{2})} \|f\|_p.$$

Proof: As $\|P_k f\|_2 \leq \|f\|_2$ it is enough to prove the above estimate when $p = 1$. Since

$$\|P_k f\|_2^2 = (P_k f, P_k f) = (P_k f, f) \leq \|P_k f\|_{p'} \|f\|_p$$

it is enough to show that

(4.4.14) $$\|P_k f\|_\infty \leq C k^{\frac{n-1}{2}} \|f\|_1.$$

To prove (4.4.14) we use the fact that $(2\pi)^n P_k = W(\varphi_k)$. This shows that P_k is an integral operator with kernel $F_k(x, y)$ given by

(4.4.15) $$F_k(x, y) = (2\pi)^{-n} \int_{\mathbb{R}^n} e^{i\xi \cdot \frac{x+y}{2}} L_k^{n-1}\left(\tfrac{1}{2}(|\xi|^2 + |x-y|^2)\right)$$
$$\times e^{-\frac{1}{4}(|\xi|^2 + |x-y|^2)} d\xi.$$

Therefore, we have the estimate

(4.4.16) $$|F_k(x, y)| \leq \int_0^\infty \left|L_k^{n-1}\left(\tfrac{1}{2} r^2\right)\right| e^{-\frac{1}{4} r^2} r^{n-1} dr.$$

Using the estimates of Lemma 1.5.4 we get

(4.4.17) $$|F_k(x, y)| \leq C k^{\frac{n-1}{2}}.$$

This proves (4.4.14) and hence the proposition. ∎

Now we can prove (ii) of Proposition 4.4.1. Taking $\alpha = -s + iy$

(4.4.18) $\|S_t(-s + iy) f\|_2^2$

$$\leq C(y) t^{-2s - n + 1} \sum_{k=0}^\infty (2k + n)^{-s - (\frac{n+1}{2})} \|P_k f\|_2^2$$

$$\leq C(y) t^{-2s - n + 1} \left(\sum_{k=0}^\infty (2k + n)^{-s - 1}\right) \|f\|_1^2$$

$$\leq C(y) t^{-2s - n + 1} \|f\|_1^2$$

as the series converges. This completes the proof of Proposition 4.4.1. ∎

We now consider another Cauchy problem for the wave operator, namely,

(4.4.19) $$\partial_t^2 u(x,t) = (-\Delta + |x|^2)u(x,t),$$

(4.4.20) $$u(x,0) = f(x), \quad \partial_t u(x,0) = 0.$$

The solution operator to the above problem is $\cos tH^{\frac{1}{2}}$ and the unique solution is given by

(4.4.21) $$u(x,t) = \sum_{k=0}^{\infty} \cos(2k+n)^{\frac{1}{2}} t P_k f(x).$$

Regarding the operator $\cos(tH^{\frac{1}{2}})$ we have the following theorem.

Theorem 4.4.2. *Assume that $\frac{2(n+1)}{n+3} < p \leq 2$ and $f \in L^p(\mathbb{R}^n)$ is such that $x_j f$ and $\frac{\partial f}{\partial x_j}$ are all in $L^p(\mathbb{R}^n)$, $j = 1, 2, \ldots, n$. Then the solution $u(x,t)$ defined in (4.4.21) satisfies*

$$\|u(\cdot, t)\|_2 \leq C \sum_{j=1}^{n} \left(\|x_j f\|_p + \left\| \frac{\partial}{\partial x_j} f \right\|_p \right)$$

where C is a constant independent of t.

Proof: As $s^{-\frac{1}{2}} \cos s = \left(\frac{\pi}{2}\right)^{\frac{1}{2}} J_{-\frac{1}{2}}(s)$ it is easier to study the operator $H^{-\frac{1}{2}} \cos(tH^{\frac{1}{2}})$ rather than $\cos(tH^{\frac{1}{2}})$. We claim that the theorem follows once we know that

(4.4.22) $$\left\| H^{-\frac{1}{2}} \cos(tH^{\frac{1}{2}}) f \right\|_2 \leq C \|f\|_p, \quad \frac{2(n+1)}{n+3} < p \leq 2.$$

To see this we can write

(4.4.23) $$\cos tH^{\frac{1}{2}} = \tfrac{1}{2}(H^{-\frac{1}{2}} \cos tH^{\frac{1}{2}}) H^{-\frac{1}{2}} \sum_{j=1}^{n}(A_j A_j^* + A_j^* A_j)$$

$$= \tfrac{1}{2}(H^{-\frac{1}{2}} \cos tH^{\frac{1}{2}}) \sum_{j=1}^{n}(H^{-\frac{1}{2}} A_j A_j^* + H^{-\frac{1}{2}} A_j^* A_j).$$

Since $H^{-\frac{1}{2}} A_j$ and $H^{-\frac{1}{2}} A_j^*$ are bounded on $L^p(\mathbb{R}^n)$, $1 < p < \infty$, it follows from (4.4.22) that

(4.4.24) $$\|\cos tH^{\frac{1}{2}} f\|_2 \leq C \sum_{j=1}^{n} \|A_j f\|_p + \|A_j^* f\|_p$$

which proves the theorem.

In order to establish (4.4.22) we would like to embed $H^{-\frac{1}{2}}\cos tH^{\frac{1}{2}}$ in an analytic family of operators as in the case of $H^{-\frac{1}{2}}\sin(tH^{\frac{1}{2}})$. This time we are led to consider several families as shown below. Suppressing the dependence on t we define the following families:

(4.4.25) $\qquad S_j(\alpha) = t^2 A_j (tH^{\frac{1}{2}})^{\alpha-\frac{n}{2}-1} J_{\frac{n}{2}-\alpha-1}(tH^{\frac{1}{2}}),$

(4.4.26) $\qquad S_j^*(\alpha) = t^2 A_j^* (tH^{\frac{1}{2}})^{\alpha-\frac{n}{2}-1} J_{\frac{n}{2}-\alpha-1}(tH^{\frac{1}{2}}).$

When $\alpha = \frac{n-1}{2}$ we have the relations

(4.4.27) $\qquad S_j\left(\frac{n-1}{2}\right) = t^{\frac{1}{2}} A_j (H^{\frac{1}{2}})^{-\frac{3}{2}} J_{-\frac{1}{2}}(tH^{\frac{1}{2}}),$

(4.4.28) $\qquad S_j^*\left(\frac{n-1}{2}\right) = t^{\frac{1}{2}} A_j^* (H^{\frac{1}{2}})^{-\frac{3}{2}} J_{-\frac{1}{2}}(tH^{\frac{1}{2}}).$

From this we have the formula

(4.4.29) $\displaystyle\sum_{j=1}^n \left\{ H^{-\frac{1}{2}} A_j^* S_j\left(\frac{n-1}{2}\right) + H^{-\frac{1}{2}} A_j S_j^*\left(\frac{n-1}{2}\right) \right\}$

$\qquad = t^{\frac{1}{2}} H^{-\frac{1}{2}} \left(\displaystyle\sum_{j=1}^n A_j^* A_j + A_j A_j^*\right) H^{-\frac{3}{4}} J_{-\frac{1}{2}}(tH^{\frac{1}{2}})$

$\qquad = 2t^{\frac{1}{2}} H^{-\frac{1}{4}} J_{-\frac{1}{2}}(tH^{\frac{1}{2}})$

$\qquad = 2\left(\dfrac{2}{\pi}\right) H^{-\frac{1}{2}} \cos(tH^{\frac{1}{2}}).$

Thus it suffices to consider the operators $S_j\left(\frac{n-1}{2}\right)$ and $S_j^*\left(\frac{n-1}{2}\right)$. We treat only $S_j\left(\frac{n-1}{2}\right)$ as the treatment of $S_j^*\left(\frac{n-1}{2}\right)$ is completely similar.

Thus we look at the analytic family $S_j(\alpha)$ defined in (4.2.25). When $\alpha = \frac{n+1}{2} + iy$

(4.4.30) $\qquad S_j\left(\dfrac{n+1}{2} + iy\right) = t(A_j H^{-\frac{1}{2}})(tH^{\frac{1}{2}})^{\frac{1}{2}+iy} J_{-\frac{3}{2}+iy}(tH^{\frac{1}{2}}).$

As both $A_j H^{-\frac{1}{2}}$ and $(tH^{\frac{1}{2}})^{\frac{1}{2}+iy} J_{-\frac{3}{2}+iy}(tH^{\frac{1}{2}})$ are bounded on $L^2(\mathbb{R}^n)$ it follows that

(4.4.31) $\qquad \left\| S_j\left(\dfrac{n+1}{2} + iy\right) f \right\|_2 \leq C(y) t \|f\|_2.$

Writing the Hermite expansion we have

$$(4.4.32) \quad S_j(\alpha)f = t(A_j H^{-\frac{1}{2}}) \sum_{k=0}^{\infty} ((2k+n)^{\frac{1}{2}}t)^{\alpha-\frac{n}{2}}$$
$$\times J_{\frac{n}{2}-\alpha-1}((2k+n)^{\frac{1}{2}}t)P_k f.$$

In view of the estimate $\|P_k f\|_2 \leq Ck^{\frac{n-1}{4}}\|f\|_1$ we immediately see that

$$(4.4.33) \quad \|S_j(-s+iy)f\|_2 \leq C(y)t^{-s-(\frac{n-1}{2})}\|f\|_1.$$

The rest is routine. By analytic interpolation we can conclude that $S_j(\frac{n-1}{2})$ are bounded from $L^p(\mathbb{R}^n)$ to $L^2(\mathbb{R}^n)$, $\frac{2(n+1)}{n+3} < p \leq 2$. This completes the proof of the Theorem. ∎

We conclude this section with the following remark. Analogues of Theorems 4.4.1 and 4.4.2 were proved by Strichartz [St1] for the wave equation $\partial_t^2 u = \Delta u$ on $\mathbb{R}^n$. More general $L^p - L^q$ estimates are known for the wave equation $\partial_t^2 u = \Delta u$. The above proofs of our theorems are modeled after the proofs given in [St1].

CHAPTER 5
HERMITE EXPANSIONS ON ℝ

In this chapter we study the almost everywhere and mean convergence of the Cesàro means $s_N^\alpha f$ for the Hermite series on the real line. We show that $\alpha = \frac{1}{6}$ is the critical index for the Cesàro summability. For $\alpha > \frac{1}{6}$, we prove almost everywhere and norm convergence of the Cesàro means to the function whenever the function is in $L^p(\mathbb{R})$, $1 \leq p < \infty$. The convergence is proved using a basic estimate for the kernel of the Cesàro means. In order to get the estimate we express the kernel as an oscillatory integral and use the method of stationary phase. By interpolating with a result of Askey and Wainger on the convergence of the partial sums we also get summability results when α is below the critical index.

5.1 Cesàro Means and the Critical Index

We are interested in the almost everywhere and mean convergence of Hermite series for functions in $L^p(\mathbb{R})$. Since the Hermite functions h_k are in $L^p(\mathbb{R})$ for all $1 \leq p \leq \infty$, the Fourier–Hermite coefficients

$$(5.1.1) \qquad \widehat{f}(k) = (f, h_k) = \int_{-\infty}^{\infty} f(x) h_k(x) dx$$

are well defined for any f in $L^p(\mathbb{R})$, $1 \leq p \leq \infty$. Therefore, to each function f in $L^p(\mathbb{R})$ we can associate the Hermite series

$$(5.1.2) \qquad f(x) = \sum_{k=0}^{\infty} \widehat{f}(k) h_k(x).$$

The above series does not converge to f in the norm unless $\frac{4}{3} < p < 4$ (see Theorem 3.1.1). As the partial sums fail to converge we are led to consider the Riesz means and the Cesàro means of the series. The Riesz means are defined in (3.1.15). In this Chapter we study the Cesàro means rather than the Riesz means. According to a theorem of Gergen [Gn], the Riesz means can be expressed in terms of the Cesàro means and vice versa. So, the norm convergence of one mean implies that of the other.

The Cesàro means, which we denote by s_N^α are defined by

$$(5.1.3) \qquad s_N^\alpha f = \frac{1}{A_N^\alpha} \sum_{k=0}^{N} A_{N-k}^\alpha \widehat{f}(k) h_k.$$

Here A_k^α are the binomial coefficients. When $\alpha = 0$ we have $s_N^0 f = S_N f$ and we know that $S_N f$ converge to f in the norm for f in $L^p(\mathbb{R})$, $\frac{4}{3} < p < 4$. When $\alpha > 0$ we expect that $s_N^\alpha f$ will converge to f for f in $L^p(\mathbb{R})$ with p lying in a bigger interval. In order to prove the possible range we proceed as follows. Let us write

$$(5.1.4) \qquad G_r(x,y) = \exp\left\{ -\frac{1}{2} \frac{1+r^2}{1-r^2}(x^2+y^2) + \frac{2rxy}{1-r^2} \right\}.$$

We prove the following generating function identity for the Cesàro means.

Theorem 5.1.1. For $0 < r < 1$ we have

$$\sum_{k=0}^{\infty} A_k^\alpha s_k^\alpha f(x) r^k = \pi^{-\frac{1}{2}}(1-r)^{-\alpha-1}(1-r^2)^{-\frac{1}{2}} \int_{-\infty}^{\infty} G_r(x,y) f(y) dy.$$

Proof: The proof is based on Mehler's formula. Recall that for $0 < r < 1$ we have

$$(5.1.5) \qquad \sum_{k=0}^{\infty} h_k(x) h_k(y) r^k = \pi^{-\frac{1}{2}}(1-r^2)^{-\frac{1}{2}} G_r(x,y).$$

Since the binomial coefficients A_k^α satisfy the generating function relation

$$(5.1.6) \qquad \sum_{k=0}^{\infty} A_k^\alpha r^k = (1-r)^{-\alpha-1},$$

we obtain by multiplying (5.1.5) and (5.1.6),

$$(5.1.7) \qquad \sum_{k=0}^{\infty} \left(\sum_{j=0}^{k} h_j(x) h_j(y) A_{k-j}^\alpha \right) r^k$$
$$= \pi^{-\frac{1}{2}}(1-r)^{-\alpha-1}(1-r^2)^{-\frac{1}{2}} G_r(x,y).$$

Multiplying both sides of (5.1.7) by $f(y)$ and integrating we obtain the theorem. ∎

HERMITE EXPANSIONS ON ℝ

Using this theorem we can express the partial sums $S_N f$ in terms of the Cesàro means $s_N^\alpha f$. In fact, multiplying the generating function

$$(5.1.8) \qquad \sum_{k=0}^{\infty} A_k^\alpha s_k^\alpha f r^k = \pi^{-\frac{1}{2}}(1-r)^{-\alpha-1}(1-r^2)^{-\frac{1}{2}} \int_{-\infty}^{\infty} G_r(x,y)f(y)dy$$

and the formula $\sum_{k=0}^{\infty} A_k^{-\alpha-1} r^k = (1-r)^\alpha$ we get the relation

$$(5.1.9) \qquad \sum_{k=0}^{\infty} \Big(\sum_{j=0}^{k} A_j^\alpha s_j^\alpha f(x) A_{k-j}^{-\alpha-1} \Big) r^k$$

$$= \pi^{-\frac{1}{2}}(1-r^2)^{-\frac{1}{2}}(1-r)^{-1} \int_{-\infty}^{\infty} G_r(x,y)f(y)dy.$$

Since the right hand side is the generating function for the partial sums we obtain

$$(5.1.10) \qquad S_N f(x) = \sum_{k=0}^{N} A_k^\alpha s_k^\alpha f(x) A_{N-k}^{-\alpha-1}.$$

Using this formula and the estimates for the L^p norms of the Hermite function we can now prove the following theorem.

Theorem 5.1.2. *If the Cesàro means s_N^α are uniformly bounded on $L^p(\mathbb{R})$ for a $p \leq 4$ then we necessarily have $p \geq 4/(6\alpha + 3)$.*

Proof: If we assume that s_N^α are uniformly bounded on $L^p(\mathbb{R})$, then in view of (5.1.10) we have

$$(5.1.11) \qquad |(f, h_N)| \, \|h_N\|_p \leq C \|f\|_p \Big(\sum_{k=0}^{N} A_k^\alpha |A_{N-k}^{-\alpha-1}| \Big).$$

As $A_k^\alpha = 0(k^\alpha)$ as $k \to \infty$, a simple calculation shows that

$$(5.1.12) \qquad |(f, h_N)| \, \|h_N\|_p \leq C N^\alpha \|f\|_p.$$

Since (5.1.12) is true for all f in $L^p(\mathbb{R})$ we obtain the estimate

$$(5.1.13) \qquad \|h_N\|_p \|h_N\|_q \leq C N^\alpha$$

where q is the index conjugate to p. If we use the estimates proved in Lemma 1.5.2 we get

$$(5.1.14) \qquad \|h_N\|_p \|h_N\|_q \geq C_1 N^{-\frac{1}{2}+\frac{2}{3p}}.$$

Therefore, (5.1.13) is valid as $N \to \infty$ only if we have $\alpha \geq -\frac{1}{2} + \frac{2}{3p}$ or $p \geq \frac{4}{6\alpha+3}$. This proves the theorem. ∎

By taking $\alpha = 0$ we infer from the above theorem that the range $\frac{4}{3} < p < 4$ in Theorem 3.1.1 is only to be expected. The theorem also proves that s_N^α are not uniformly bounded on $L^1(\mathbb{R})$ unless $\alpha \geq \frac{1}{6}$. At this point we define the critical index α_{cr} for the Cesàro summability to be the smallest index with the property that $\alpha > \alpha_{cr}$ implies uniform boundedness of s_N^α on $L^1(\mathbb{R})$ (and hence on all $L^p(\mathbb{R})$, $1 \leq p \leq \infty$). Thus we observe that $\alpha_{cr} \geq \frac{1}{6}$ for the Hermite series on $\mathbb{R}$. The main theorem of this chapter is to prove that $\alpha_{cr} = \frac{1}{6}$ indeed. We conclude this section with the following remarks concerning the critical index. In Chapter 3 we proved that the critical index for the summability of Hermite expansions on $\mathbb{R}^n$, $n \geq 2$ is $(n-1)/2$. We also note that in the case of Fourier series, the critical index is 0 and for the Laguerre expansion on $[0, \infty)$ the critical index is $\frac{1}{2}$. Thus, we see that the one dimensional Hermite series has summability properties which are different from those of Fourier or Laguerre series. The fact that $\alpha_{cr} = \frac{1}{6}$ is in agreement with a conjecture of Lorch which states that if the partial sum operators S_N have norm $\|S_N\|_1$ growing like N^ν as $N \to \infty$ on L^1 spaces then $\alpha_{cr} = \nu$. This conjecture holds good for many expansions including Fourier and Laguerre series.

5.2 An Expression for the Cesàro Kernel and the Basic Estimates

Consider the kernel $s_N^\alpha(x,y)$ of the Cesàro means s_N^α defined by

(5.2.1) $$s_N^\alpha(x,y) = \frac{1}{A_N^\alpha} \sum_{k=0}^{N} A_{N-k}^\alpha h_k(x) h_k(y).$$

Our aim in this section is to express the Cesàro kernel as a sum of certain oscillatory integrals. The oscillatory integrals are estimated in the next section.

We start with the generating function identity (5.1.7) proved in the previous section. Since (5.1.7) is true even when r is complex with $|r| < 1$ we replace r by re^{-2it} obtaining

(5.2.2) $$\sum_{k=0}^{\infty} A_k^\alpha s_k^\alpha(x,y) r^k e^{-2kit}$$
$$= \pi^{-\frac{1}{2}} (1 - re^{-2it})^{-\alpha-1} (1 - r^2 e^{-4it})^{-\frac{1}{2}} e^{B_r(t,x,y)}$$

where we have written

(5.2.3) $$B_r(t,x,y) = \exp\left(-\frac{1}{2} \frac{1 + r^2 e^{-4it}}{1 - r^2 e^{-4it}}(x^2 + y^2) + \frac{2re^{-2it}}{1 - r^2 e^{-4it}} xy\right).$$

Multiplying both sides of (5.2.2) by e^{2Nit} and integrating from $-\frac{\pi}{2}$ to $\frac{\pi}{2}$ we obtain

$$(5.2.4) \quad r^N A_N^\alpha s_N^\alpha(x,y) = \pi^{-\frac{3}{2}} \int_{-\frac{\pi}{2}}^{\frac{\pi}{2}} (1-re^{-2it})^{-\alpha-1}$$
$$\times (1-r^2 e^{-4it})^{-\frac{1}{2}} e^{2Nit} e^{B_r(t,x,y)} dt.$$

We express the right hand side as a sum of three integrals in the following way.

We choose a cut off function θ such that $\theta(t) = 0$ for $|t| \leq \frac{1}{2}$ and $\theta(t) = 1$ for $|t| > 1$. We let $R = 2N + \alpha + 2$ and define

$$(5.2.5) \quad I_r = \int_{|t| \leq \frac{\pi}{4}} (1-\theta(Rt))(1-re^{-2it})^{-\alpha-1}$$
$$\times (1-r^2 e^{-4it})^{-\frac{1}{2}} e^{2Nit} e^{B_r(t,x,y)} dt,$$

$$(5.2.6) \quad J_r = \int_{|t| \leq \frac{\pi}{4}} \theta(Rt)(1-re^{-2it})^{-\alpha-1}$$
$$\times (1-r^2 e^{-4it})^{-\frac{1}{2}} e^{2Nit} e^{B_r(t,x,y)} dt,$$

$$(5.2.7) \quad K_r = \int_{|t| > \frac{\pi}{4}} (1-re^{-2it})^{-\alpha-1}$$
$$\times (1-r^2 e^{-4it})^{-\frac{1}{2}} e^{2Nit} e^{B_r(t,x,y)} dt.$$

In the integral defining K_r we can make a change variable to write it in the form

$$(5.2.8) \quad K_r = (-1)^N \int_{|t| \leq \frac{\pi}{4}} (1+re^{-2it})^{-\alpha-1}$$
$$\times (1-r^2 e^{-4it})^{-\frac{1}{2}} e^{2Nit} e^{B_r(t,x,-y)} dt.$$

Thus we have obtained the expression

$$(5.2.9) \quad s_N^\alpha(x,y) = \lim_{r \to 1} \frac{\pi^{-\frac{3}{2}}}{A_N^\alpha} \{I_r + J_r + K_r\}.$$

We claim that we can pass to the limit under the integral sign in J_r and K_r. In fact, a simple calculation shows that

$$(5.2.10) \quad \lim_{r \to 1} B_r(t,x,y) = i\varphi(t,x,y),$$

(5.2.11) $$\varphi(t,x,y) = -xy\operatorname{cosec} 2t + \tfrac{1}{2}(x^2+y^2)\cot 2t.$$

Also we have

(5.2.12)
$$\lim_{r\to 1}(1-r^2 e^{-4it})^{-\frac{1}{2}} = (2i)^{-\frac{1}{2}}e^{it}(\sin 2t)^{-\frac{1}{2}},$$
$$\lim_{r\to 1}(1-re^{-2it})^{-\alpha-1} = (2i)^{-\alpha-1}e^{(\alpha+1)it}(\sin t)^{-\alpha-1}$$
$$\lim_{r\to 1}(1+re^{-2it})^{-\alpha-1} = 2^{-\alpha-1}e^{i(\alpha+1)t}(\cos t)^{-\alpha-1}.$$

In view of these relations it is easy to prove that in J_r and K_r we can pass to the limit. Let us write $\lim_{r\to 1} I_r = I$, $\lim_{r\to 1} J_r = J$ and $\lim_{r\to 1} K_r = K$ so that

(5.2.13) $$J = (2i)^{-(\alpha+\frac{3}{2})} \int_{|t|\le\frac{\pi}{4}} \theta(Rt)(\sin t)^{-\alpha-1}(\sin 2t)^{-\frac{1}{2}} e^{i\varphi(t,x,y)} dt$$

and

(5.2.14) $$K = i^{-\frac{1}{2}} 2^{-(\alpha+\frac{3}{2})} \int_{|t|\le\frac{\pi}{4}} (\sin 2t)^{-\frac{1}{2}}(\cos t)^{-\alpha-1} e^{i\varphi(t,x,-y)} dt.$$

Thus we have obtained the following expression for the Cesàro kernel.

Proposition 5.2.1. *With some constants C_1, C_2 and C_3 we have the formula*

(5.2.15) $$A_N^\alpha s_N^\alpha(x,y) = C_1 I + C_2 J + C_3 K.$$

Having obtained a usable expression for the Cesàro kernel, we now turn our attention to get good estimates. We start with a rough estimate for the kernel, namely,

Lemma 5.2.1. *There is a constant C independent of N, x and y such that*

(5.2.16) $$|s_N^\alpha(x,y)| \le CN^{\frac{1}{2}}.$$

Proof: We observe that, as $A_k^\alpha = \dfrac{\Gamma(k+\alpha+1)}{\Gamma(k+1)\Gamma(\alpha+1)} = O(k^\alpha)$,

(5.2.17) $$|s_N^\alpha(x,y)| \le \sum_{k=0}^{N} |h_k(x)|\,|h_k(y)|.$$

In view of the Cauchy–Schwarz inequality it is enough to prove

(5.2.18) $$\sum_{k=0}^{N}(h_k(x))^2 \leq CN^{\frac{1}{2}},$$

which has been already proved in Lemma 3.2.1. ∎

For the integrals J and K it is easy to prove the estimates

(5.2.19) $$|J| \leq CN^{\alpha+\frac{1}{2}}, \quad |K| \leq C.$$

In fact, a simple integration proves these estimates as $R = (2N+2+\alpha)$. Therefore, (5.2.19) and (5.2.16) give us the estimate

(5.2.20) $$\frac{1}{A_N^\alpha}|I| \leq CN^{\frac{1}{2}}.$$

Regarding I, we now establish the following:

Proposition 5.2.2. Assume that $0 \leq \alpha < \frac{1}{2}$. Then there is a constant C independent N, x and y such that

(5.2.21) $$\frac{1}{A_N^\alpha}|I| \leq CN^{\frac{1}{2}}(1+N^{\frac{1}{2}}|x-y|)^{-2}.$$

Proof: Since we have the estimate (5.2.20) we need to prove that

(5.2.22) $$|I| \leq CN^{\alpha-\frac{1}{2}}|x-y|^{-2}, \quad \text{for } |x-y| \geq N^{-\frac{1}{2}}.$$

To prove this, we perform an integration by parts in I_r and then pass to the limit as $r \to 1$. Recall that I_r is given by

(5.2.23) $$I_r = \int g_r(t) e^{B_r(t,x,y)} dt$$

where the function $g_r(t)$, given by

(5.2.24) $$g_r(t) = (1-\theta(Rt))(1-r^2 e^{-4it})^{-\frac{1}{2}}(1-re^{-2it})^{-\alpha-1} e^{2Nit}$$

is supported in $|t| \leq R^{-1}$. Let us write

(5.2.25) $$g_r(t) e^{B_r(t)} = \frac{d}{dt}\{g_r(t) B_r'(t)^{-1} e^{B_r(t)}\}$$
$$- \frac{d}{dt}\{g_r(t) B_r'(t)^{-1}\} e^{B_r(t)}$$

where $B_r(t) = B_r(t,x,y)$. Integrating by parts in I_r we get

$$(5.2.26) \qquad I_r = -\int \frac{d}{dt}\{g_r(t)B_r'(t)^{-1}\}e^{B_r(t)}dt.$$

As r tends to 1, $B_r'(t)$ converges to $i\varphi'(t,x,y)$. A simple calculation shows that

$$(5.2.27) \qquad -\varphi'(t,x,y) = \operatorname{cosec}^2 2t\{4xy\sin^2 t + |x-y|^2\}.$$

Therefore, it is easily seen that

$$(5.2.28) \qquad \lim_{r\to 1} g_r(t) = (2i)^{-(\alpha+\frac{3}{2})}(\sin 2t)^{-\frac{1}{2}}(\sin t)^{-\alpha-1}$$
$$\times e^{iRt}(i\varphi'(t))^{-1}(1-\theta(Rt)).$$

As we are integrating over $|t| \leq R^{-1}$ we see that

$$(5.2.29) \qquad -\varphi'(t)\sin^2 2t \geq \tfrac{1}{2}|x-y|^2.$$

So, we have the estimate

$$(5.2.30) \qquad \left|\lim_{r\to 1} g_r(t)B_r'(t)^{-1}\right| \leq Ct^{-\alpha+\frac{1}{2}}|x-y|^{-2}.$$

We claim that we also have

$$(5.2.31) \qquad \left|\lim_{r\to 1}\frac{d}{dt}\{g_r(t)B_r'(t)^{-1}\}\right| \leq Ct^{-\alpha-\frac{1}{2}}|x-y|^{-2}.$$

When the derivative falls on the factor

$$(5.2.32) \qquad (\sin 2t)^{-\frac{1}{2}}(\sin t)^{-\alpha-1}e^{iRt}(1-\theta(Rt))(\varphi'(t))^{-1}$$

we claim that it brings down a factor of t^{-1}. This is clear if the derivative falls on the sine functions. When the derivative falls on e^{iRt} or $(1-\theta(Rt))$ it brings down R which is bounded by t^{-1} as $|t| \leq R^{-1}$ on the range of integration. When the derivative falls on $(\varphi'(t))^{-1}$ we get $\varphi''(t)(\varphi'(t))^{-2}$. We claim that

$$(5.2.33) \qquad |\varphi''(t)(\varphi'(t))^{-2}| \leq C(\sin 2t)|x-y|^{-2}.$$

This claim follows at once if we show that

$$(5.2.34) \qquad |\varphi''(t)\sin 2t| \leq 4|\varphi'(t)|, \quad 0 \leq t \leq \frac{\pi}{4}.$$

In order to prove (5.2.34) we calculate the second derivative to be

(5.2.35) $\varphi''(t)\sin^3 2t = 4\bigl(-4xy\sin^4 t + |x-y|^2\cos 2t\bigr).$

Therefore, when $xy \geq 0$ it is clear that (5.2.34) is true as $|\varphi'(t)\sin^2 2t| = 4xy\sin^2 t + |x-y|^2$. When $xy < 0$, (5.2.34) will follow if

$$-4xy\sin^4 t + |x-y|^2\cos 2t \leq 4xy\sin^2 t + |x-y|^2.$$

In other words we need to check if $-4xy\sin^2 t(1+\sin^2 t) \leq 2\sin^2 t|x-y|^2$, which is obviously true. Thus we have the estimate (5.2.32). If we assume that $\alpha < \tfrac{1}{2}$ then it is clear that we can pass to the limit under the integral sign and also we have the estimate

$$|I| \leq C|x-y|^{-2}\int_0^{R^{-1}} t^{-\alpha-\tfrac{1}{2}}\,dt \leq CR^{\alpha-\tfrac{1}{2}}|x-y|^{-2}.$$

This completes the proof of the proposition. ∎

We estimate the oscillatory integrals J and K in the next section. We prove

(5.2.36) $|J| \leq CN^{\alpha+\tfrac{1}{2}}(1+N^{\tfrac{1}{2}}|x-y|)^{-\alpha-\tfrac{5}{6}}$

(5.2.37) $|K| \leq CN^{\alpha+\tfrac{1}{2}}(1+N^{\tfrac{1}{2}}|x+y|)^{-\alpha-\tfrac{5}{6}}$

under the assumption that $\tfrac{1}{6} < \alpha < \tfrac{1}{2}$. These two estimates together with Proposition 5.2.2 will give the basic estimate for the Cesàro kernel.

5.3 Estimating the Oscillatory Integrals

In this section we establish the estimates (5.2.36) and (5.2.37) for the oscillatory integrals J and K. Together with Proposition 5.2.2 the estimates for J and K will then prove the following estimate for the Cesàro kernel.

Theorem 5.3.1. *Assume that $\tfrac{1}{6} < \alpha < \tfrac{1}{2}$. Then we have the estimate*

$$|s_N^\alpha(x,y)| \leq CN^{\tfrac{1}{2}}\bigl\{(1+N^{\tfrac{1}{2}}|x-y|)^{-\alpha-\tfrac{5}{6}} + (1+N^{\tfrac{1}{2}}|x+y|)^{-\alpha-\tfrac{5}{6}}\bigr\}.$$

We now proceed to the estimation of J and K. Since the phase function for the integral K is $Rt + \varphi(t, x, -y)$ it is enough to consider J.

The estimate for K is obtained from that of J by simply replacing y by $-y$. Since we already have the trivial estimate $|J| \leq CN^{\alpha+\frac{1}{2}}$ we need only to prove the estimate

$$|J| \leq CN^{\frac{\alpha}{2}+\frac{1}{12}}|x-y|^{-\alpha-\frac{5}{6}}$$

valid in the region $|x-y| \geq 2R^{-\frac{1}{2}}$. We replace x and y by $R^{\frac{1}{2}}x$ and $R^{\frac{1}{2}}y$ and consider the integral

$$(5.3.1) \qquad A = \int_0^{\frac{\pi}{4}} \theta(Rt)(\sin 2t)^{-\frac{1}{2}}(\sin t)^{-\alpha-1}e^{iR(t+\varphi(t))}dt.$$

For this integral we establish the following estimate.

Proposition 5.3.1. *Under the assumption that $\frac{1}{6} < \alpha < \frac{1}{2}$ and $|x-y| \geq 2R^{-1}$ one has the estimate*

$$|A| \leq CR^{-\frac{1}{3}}|x-y|^{-\alpha-\frac{5}{6}}.$$

Let us define functions $\omega_R(t)$ and $g_R(t)$ by the equations

$$(5.3.2) \qquad \omega_R(t) = \theta(Rt)\left(\frac{t}{\sin 2t}\right)^{\frac{1}{2}}\left(\frac{t}{\sin t}\right)^{\alpha+1},$$

$$(5.3.3) \qquad g_R(t) = \omega_R(t)t^{-\alpha-\frac{3}{2}}.$$

We also set $\psi(t) = t + \varphi(t)$ so that

$$(5.3.4) \qquad A = \int_0^{\frac{\pi}{4}} g_R(t)e^{iR\psi(t)}dt.$$

We observe that the function $\omega_R(t)$ satisfies

$$(5.3.5) \qquad |\omega_R(t)| \leq C_1, \ |\omega_R'(t)| \leq C_2 t^{-1} \quad \text{for } 0 \leq t \leq \frac{\pi}{4}.$$

In order to estimate A we need the following oscillatory integral theorem called the Van der Corput lemma. The proof is taken from Stein [S3].

Theorem 5.3.2. *Suppose ψ is real valued and smooth in $[a,b]$. If $|\psi^{(k)}(t)| \geq 1$, then*

$$\left| \int_a^b \varphi(t) e^{iR\psi(t)} dt \right| \leq CR^{-\frac{1}{k}} \left\{ |\varphi(b)| + \int_a^b |\varphi'(t)| dt \right\}$$

holds when i) $k \geq 2$ *or* ii) $k = 1$, *if in addition $\psi'(t)$ is monotonic.*

Proof: We first consider the case $\varphi(t) \equiv 1$. Let D denote the differential operator $Df = \dfrac{1}{iR\psi'(t)} \dfrac{df}{dt}$ and D^* be its transpose, $D^* f = \dfrac{d}{dt}\left(\dfrac{f}{iR\psi'}\right)$. Then clearly, $D(e^{iR\psi}) = e^{iR\psi}$. We consider the case $k = 1$ first. We have

$$(5.3.6) \qquad \int_a^b e^{iR\psi(t)} dt = \int_a^b D(e^{iR\psi}) dt$$

$$= \left. \frac{e^{iR\psi(t)}}{iR\psi'(t)} \right|_a^b - \int_a^b e^{iR\psi(t)} D^*(1) dt.$$

The boundary terms are majorized by $2/R$, while

$$(5.3.7) \qquad \left| \int_a^b e^{iR\psi(t)} D^*(1) dt \right| = \left| \int_a^b e^{iR\psi(t)} (iR)^{-1} \frac{d}{dt}\left(\frac{1}{\psi'(t)}\right) dt \right|$$

$$\leq \frac{1}{R} \int_a^b \left| \frac{d}{dt}\left(\frac{1}{\psi'}\right) \right| dt = \frac{1}{R} \left| \int_a^b \frac{d}{dt}\left(\frac{1}{\psi'}\right) dt \right|$$

by the monotonicity of ψ'. The last expression equals $R^{-1} \left| \frac{1}{\psi'(b)} - \frac{1}{\psi'(a)} \right|$ which is dominated by $2/R$. This proves the proposition when $k = 1$.

We now prove the case $k > 1$ by induction. Let us assume that the case k is known and assume that $|\psi^{(k+1)}(t)| \geq 1$. Let $t = c$ be the point in $[a,b]$ where $\psi^{(k)}(t)$ takes its minimum value. If $\psi^{(k)}(c) = 0$ then outside the interval $(c - \delta, c + \delta)$ we have that $|\psi^{(k)}(t)| \geq \delta$. Write

$$(5.3.8) \qquad \int_a^b e^{iR\psi(t)} dt = \int_a^{c-\delta} + \int_{c-\delta}^{c+\delta} + \int_{c+\delta}^b e^{iR\psi(t)} dt.$$

By the previous case the first and third integral gives the estimate $c_k(R\delta)^{-\frac{1}{k}}$ while the middle integral is bounded by 2δ. Thus

$$(5.3.9) \qquad \left| \int_a^b e^{iR\psi} dt \right| \leq 2\delta + c_k(R\delta)^{-\frac{1}{k}}.$$

If $\psi^{(k)}(c) \neq 0$ then c will be an end point and a similar argument will show that (5.3.9) is true. In either situation the case $k+1$ follows by taking $\delta = R^{-\frac{1}{k+1}}$.

Now to prove the general case we consider

$$\int_a^b e^{iR\psi(t)}\varphi(t)dt = \int_a^b \varphi(t)d\left(\int_a^t e^{iR\psi(s)}ds\right).$$

Integrating by parts and using the estimate

$$\left|\int_a^b e^{iR\psi(t)}dt\right| \leq C_k R^{-\frac{1}{k}}$$

we obtain the proposition. ∎

We choose a positive number $\delta \leq \frac{1}{4}$ to be fixed later and estimate A in the regions $2R^{-1} \leq |x-y| \leq 2\delta$ and $2\delta \leq |x-y|$ separately. Our main tool in estimating A is Theorem 5.3.2 and hence we need to get lower bounds for the derivatives of the phase function $\psi(t)$.

Let $a = (x^2 + y^2)$, $b = xy$ and $\lambda = \cos 2t$. Then a simple calculation gives us

(5.3.10) $\qquad -\psi'(t)\sin^2 2t = (a^2 - 2b\lambda + \lambda^2 - 1),$

(5.3.11) $\qquad \psi''(t)\sin^3 2t = 4(a^2\lambda - b\lambda^2 - b),$

(5.3.12) $\qquad -\psi'''(t)\sin^4 2t = 8(2a^2\lambda^2 - b\lambda^3 - 5b\lambda + a^2).$

We first obtain estimates for the third derivative of ψ.

Lemma 5.3.1.

(5.3.13) $\qquad |\psi'''(t)| \geq 4|x-y|^2, \quad \text{for } 0 \leq t \leq \frac{\pi}{4},$

(5.3.14) $\qquad |\psi'''(t)| \geq b, \quad \text{for } 0 \leq t \leq \frac{\pi}{4} \quad \text{if } b \geq 0.$

Proof: To prove (5.3.13) we need to check if

$$2(2a^2\lambda^2 - b\lambda^3 - 5b\lambda + a^2) \geq (a^2 - 2b).$$

When b is negative the above is clearly true. When b is positive we can actually prove

$$(2a^2\lambda^2 - b\lambda^3 - 5b\lambda + a^2) \geq (a^2 - 2b).$$

In fact, as $a^2 \geq 2b$ it is enough to check if $G(\lambda) = (4\lambda^2 - \lambda^3 - 5\lambda + 2)$ is nonnegative which is true as it can be easily verified that $G(\lambda)$ attains a minimum of 0 at $\lambda = 1$. To prove (2.3.14) we need to check if

$$8(2a^2\lambda^2 - b\lambda^3 - 5b\lambda + a^2)(1-\lambda^2)^{-2} \geq b.$$

Again, as $a^2 \geq 2b$ we need to check if

$$8(4\lambda^2 - \lambda^3 - 5\lambda + 2) \geq (1-\lambda^2)^2.$$

But $(4\lambda^2 - \lambda^3 - 5\lambda + 2) = (2-\lambda)(1-\lambda)^2$ and hence the above inequality reduces to $8(2-\lambda) \geq (1+\lambda)^2$ which is clearly true as $0 \leq \lambda \leq 1$. ∎

We next proceed to get lower bounds for the second derivative $\psi''(t)$ near the stationary point, namely, near the point at which $\psi'(t)$ vanishes. We consider only those points x and y for which $x^2 + y^2 \leq \frac{1}{2}$. A calculation shows that the stationary point t_1 is given by $\cos 2t_1 = b + m$ where $m^2 = (1-x^2)(1-y^2)$. We also calculate $\psi''(t_1)$ to be

(5.3.15) $$\psi''(t_1) = 4m \operatorname{cosec} 2t_1 \geq 2 \operatorname{cosec} 2t_1$$

as $m \geq \frac{1}{2}$.

Lemma 5.3.2. *Assume that $|x-y| \leq 2\delta$ and $x^2 + y^2 \leq \frac{1}{2}$. Then the lower bound $|\psi''(t)| \geq \operatorname{cosec} 2t_1$, is valid under any of the following two conditions:*

i) $\frac{1}{4}|x-y| \leq t \leq (t_1 + \frac{1}{20}\sin 2t_1), \quad xy \leq 0$

ii) $(t_1 - \frac{1}{20}\sin 2t_1) \leq t \leq (t_1 + \frac{1}{20}\sin 2t_1), \quad xy > 0.$

Proof: First we consider the case $xy \leq 0$. From the formula (2.3.12) it is clear that $|\psi'''(t)|$ is decreasing. Therefore, mean value theorem gives

$$|\psi''(t) - \psi''(t_1)| \leq |\psi'''(t_1)|(t-t_1)$$

in the interval $t_1 \leq t \leq \frac{\pi}{4}$. We also have

(5.3.16) $$|\psi'''(t)| = 6 \cot 2t \psi''(t) + 8 \operatorname{cosec}^2 2t(a^2 - 2b \cos 2t).$$

Since $\psi''(t_1) = 4m \operatorname{cosec} 2t_1 \geq 2 \operatorname{cosec} 2t_1$ we see that

$$|\psi'''(t_1)| \leq 10 \operatorname{cosec} 2t_1 \psi''(t_1)$$

so that

$$\psi''(t_1) - \psi''(t) \leq 10 \operatorname{cosec} 2t_1 \psi''(t_1)(t-t_1).$$

Therefore, if we take $t_1 \leq t \leq t_1 + \frac{1}{20}\sin 2t_1$ we have

(5.3.17) $$\psi''(t_1) - \psi''(t) \leq \tfrac{1}{2}\psi''(t_1),$$

which immediately gives the lower bound

$$\psi''(t) \geq \tfrac{1}{2}\psi''(t_1) \geq \operatorname{cosec} 2t_1.$$

Also, $\psi''(t)$ is decreasing in the interval $\tfrac{1}{4}|x-y| \leq t \leq t_1$ as $\psi'''(t)$ is negative. Therefore $\psi''(t) \geq \psi''(t_1) \geq \operatorname{cosec} 2t_1$ follows in this interval also. This proves the lemma when $xy \leq 0$. The case $xy > 0$ is similar; the details are left to the interested reader. ∎

Finally we get lower bounds for the first derivative $\psi'(t)$.

Lemma 5.3.3. *Let $|x-y| \leq 2\delta$ as before. Under any of the following assumptions*

i) $x^2 + y^2 \leq \tfrac{1}{2}$, $xy \leq 0$ and $t \geq t_1 + \tfrac{1}{20}\sin 2t_1$,
ii) $x^2 + y^2 \leq \tfrac{1}{2}$, $xy > 0$ and $t \leq t_1 - \tfrac{1}{20}\sin 2t_1$,
iii) $x^2 + y^2 \leq \tfrac{1}{2}$, $xy > 0$ and $t \geq t_1 + \tfrac{1}{20}\sin 2t_1$,
iv) $x^2 + y^2 > 4$, $xy > 0$ and $\tfrac{1}{4}|x-y| \leq t \leq \tfrac{\pi}{4}$,

the estimate $|\psi'(t)| \geq \tfrac{1}{20}$ is valid.

Proof: Again we first consider the case $x^2 + y^2 \leq \tfrac{1}{2}$ and $xy \leq 0$. Since $\psi'(t_1) = 0$ Taylor's theorem applied to $\psi'(t)$ gives

$$\psi'(t) = (t-t_1)\psi''(t_1) + \int_{t_1}^{t}(t-s)\psi'''(s)\,ds$$
$$= (t-t_1)\psi''(t) + \int_{t_1}^{t}(s-t_1)(-\psi'''(s))\,ds.$$

Since $\psi'''(s)$ is negative one gets

(5.3.18) $$\psi'(t) \geq (t-t_1)\psi''(t)$$

for $t \geq t_1$. From the previous lemma we have

$$\psi''(t) \geq \tfrac{1}{2}\psi''(t_1)$$

in the interval $t_1 \leq t \leq t_1 + \tfrac{1}{20}\sin 2t_1$. Hence we have, from (5.3.18)

(5.3.19) $$\psi'(t) \geq \tfrac{1}{2}(t-t_1)\psi''(t_1), \quad t_1 \leq t \leq t_1 + \tfrac{1}{20}\sin 2t_1.$$

HERMITE EXPANSIONS ON ℝ   125

Since $\psi''(t) > 0$, $\psi'(t)$ is increasing and hence for $t \geq t_1 + \frac{1}{20}\sin 2t_1$

(5.3.20) $\qquad \psi'(t) \geq \frac{1}{2} \cdot \frac{1}{20}\sin 2t_1 \cdot \psi''(t_1) \geq \frac{1}{20}.$

In the case when $xy > 0$, $\psi'(t)$ increases up to the point t_0 given by $\cos 2t_0 = \frac{x}{y}$ (if $x < y$) or $\cos 2t_0 = \frac{y}{x}$ (if $y < x$) at which points $\psi''(t_0) = 0$ and then decreases. In the interval $t_1 + \frac{1}{20}\sin 2t_1 \leq t \leq t_0$ we can prove $|\psi'(t)| \geq \frac{1}{20}$ as above. In the interval $t_0 \leq t \leq \frac{\pi}{4}$, as $\psi'(t)$ decreases we get $\psi'(t) \geq \psi'(\frac{\pi}{4}) = (1 - a^2) \geq \frac{1}{2}$. It remains to consider $t \leq t_1 - \frac{1}{20}\sin 2t_1$. In this case

$$\psi'(t) = (t - t_1)\psi''(t_1) + \int_{t_1}^{t}(t - s)\psi'''(s)ds$$

gives, for $t_1 - \frac{1}{20}\sin 2t_1 \leq t \leq t_1$,

$$-\psi'(t) = (t_1 - t)\psi''(t_1) + \int_{t}^{t_1}(s - t)(-\psi'''(s))ds$$
$$\geq (t_1 - t)\psi''(t_1).$$

Again, as $-\psi'(t)$ is decreasing in the interval $\frac{1}{4}|x - y| \leq t \leq t_1 - \frac{1}{20}\sin 2t_1$ we obtain

$$-\psi'(t) \geq \frac{1}{20}\sin 2t_1 \cdot 2\operatorname{cosec} 2t_1 \geq \frac{1}{10}.$$

Finally, when $x^2 + y^2 > 4$, $\psi'(t)$ attains maximum at t_0. A calculation shows that this maximum is $(1 - x^2)$ or $(1 - y^2)$ depending on whether $y < x$ or $x < y$. Since $x^2 + y^2 > 4$ in either case $\psi'(t) \leq -1$. Hence $|\psi'(t)| \geq 1$. ∎

We are ready to estimate the oscillatory integral A. First we consider the case when $|x - y| \geq 2\delta$.

Lemma 5.3.4. *The estimate*

$$|A| \leq CR^{-\frac{1}{3}}|x - y|^{-\alpha - \frac{2}{6}}$$

is valid when $|x - y| \geq 2\delta$.

Proof: We want to apply the oscillatory integral theorem with $k = 3$. Before that we need to perform an integration by parts. We write

$$A = -iR^{-1}\int_{0}^{\frac{\pi}{4}} g_R(t)(\varphi'(t))^{-1}e^{iRt}d(e^{iR\varphi}).$$

Integrating by parts we see that there are two integrals to estimate, namely

$$(5.3.21) \qquad A_1 = iR^{-1} \int_0^{\frac{\pi}{4}} \frac{d}{dt}\{g_R(t)(\varphi'(t))^{-1}\} e^{iR\psi(t)} dt$$

and

$$(5.3.22) \qquad A_2 = -\int_0^{\frac{\pi}{4}} g_R(t)(\varphi'(t))^{-1} e^{iR\psi(t)} dt.$$

If we use the estimates (5.2.29) and (5.2.33) we immediately get

$$|A_1| \leq CR^{-1}|x-y|^{-2} \int_{R^{-1}}^{\frac{\pi}{4}} t^{-\alpha-\frac{1}{2}} dt \leq CR^{-1}|x-y|^{-2}$$

since $\alpha < \frac{1}{2}$. This gives the estimate $|A| \leq CR^{-\frac{1}{3}}|x-y|^{-\alpha-\frac{5}{6}}$ as $|x-y| \geq 2\delta$.

To estimate A_2 we use the lower bound $|\psi'''(t)| \geq 4|x-y|^2$ proved in Lemma 5.3.1. Applying the oscillatory integral theorem and using (5.2.29) and (5.2.33) as above we get

$$|A_2| \leq CR^{-\frac{1}{3}}|x-y|^{-2-\frac{2}{3}} \leq CR^{-\frac{1}{3}}|x-y|^{-\alpha-\frac{5}{6}}.$$

This completes the proof of the lemma. ∎

We now consider the region $2R^{-1} \leq |x-y| \leq 2\delta$. We split the integral A into two parts. Let $\beta = \frac{1}{2}|x-y|$ and define

$$(5.3.23) \qquad B_0 = \int_0^{\frac{1}{2}\beta} g_R(t) e^{iR\psi(t)} dt,$$

and

$$(5.3.24) \qquad B = \int_{\frac{1}{2}\beta}^{\frac{\pi}{4}} g_R(t) e^{iR\psi(t)} dt.$$

We can easily get an estimate for B_0. In fact, a simple calculation shows that $-\psi'(t)\sin^2 2t \geq \frac{1}{4}|x-y|^2$ for $0 \leq t \leq \frac{1}{2}\beta$ and an integration by parts will give the estimate for B_0. So, we concentrate our attention to B.

Lemma 5.3.5. *The estimate*
$$|B| \leq CR^{-\frac{1}{3}}|x-y|^{-\alpha-\frac{5}{6}}$$
is valid when $|x-y| \leq 2\delta$, *provided* $x^2 + y^2 \leq \frac{1}{2}$.

Proof: We have two cases to consider.

Case (i). $xy \leq 0$ In this case we split the integral into two parts:
$$B = \int_{\beta/2}^{t_1 + \frac{1}{20}\sin 2t_1} + \int_{t_1 + \frac{1}{20}\sin 2t_1}^{\frac{\pi}{4}} = B_1 + B_2, \quad \text{say}.$$

For the first integral we use the estimate of Lemma 5.3.2, i.e., $|\psi''(t)| \geq \operatorname{cosec} 2t_1$. Application of the oscillatory integral theorem gives
$$|B_1| \leq CR^{-\frac{1}{2}}(\sin 2t_1)^{\frac{1}{2}} \left\{ t_1^{-\alpha-\frac{3}{2}} + \int_{\beta/2}^{2t_1} t^{-\alpha-\frac{5}{2}} dt \right\}$$
$$\leq CR^{-\frac{1}{2}} \left\{ (\sin 2t_1)^{-\alpha-1} + |x-y|^{-\alpha-\frac{3}{2}}(\sin 2t_1)^{\frac{1}{2}} \right\}.$$

Now as xy is negative and t_1 is the stationary point satisfying $\sin^2 2t_1 = a^2 - 2b\cos 2t_1$ we have $\frac{1}{2}|x-y|^2 \leq a^2 \leq \sin^2 2t_1 \leq a^2 - 2b = |x-y|^2$. Therefore, we have
$$|B_1| \leq CR^{-\frac{1}{2}}|x-y|^{-\alpha-1} \leq CR^{-\frac{1}{3}}|x-y|^{-\alpha-\frac{5}{6}}.$$

The estimation of B_2 is similar; we use the lower bound (i) of Lemma 5.3.3.

Case (ii). $xy > 0$ In this case split the integral into four parts corresponding to the intervals $\beta/2 \leq t \leq t_1 - \frac{1}{20}\sin 2t_1$, $t_1 - \frac{1}{20}\sin 2t_1 \leq t \leq t_1 + \frac{1}{20}\sin 2t_1$, $t_1 + \frac{1}{20}\sin Rt_1 \leq t \leq t_0$ and $t_0 \leq t \leq \frac{\pi}{4}$. For the first, third and last parts of the integrals we use Lemma 5.3.3 and for the second part we use Lemma 5.3.2. Again observing that $\sin^2 2t_1 \geq |x-y|^2$ we complete the proof. ∎

In the previous lemma we estimated B when $x^2 + y^2 \leq \frac{1}{2}$. It remains to consider $x^2 + y^2 \geq \frac{1}{2}$. Since we are assuming that $|x-y| \leq 2\delta$, $\delta \leq \frac{1}{4}$ it follows that $xy > 0$ when $x^2 + y^2 \geq \frac{1}{2}$. When $x^2 + y^2 > 4$ we have the lower bound $|\psi'(t)| \geq 1$ from Lemma 5.3.3 and hence estimate for B follows immediately. Thus we need to consider B when $\frac{1}{2} \leq x^2 + y^2 \leq 4$. Again we make one more splitting of the integral. Write $B = F_0 + F$ where (with $\rho = 1$ or $\frac{1}{3}$)

(5.3.25) $$F_0 = \int_{\rho\beta^{3/5}}^{\frac{\pi}{4}} g_R(t) e^{iR\psi(t)}.$$

We first estimate F_0.

Lemma 5.3.6. *The estimate*
$$|F_0| \leq CR^{-\frac{1}{3}}|x-y|^{-\alpha-5/6}$$
is valid under the assumption that $\frac{1}{2} \leq x^2 + y^2 \leq 4$.

Proof: We use the estimate $|\psi'''(t)| \geq xy$ of Lemma 5.3.1. Since $|x-y| \leq \frac{1}{2}$ and $x^2 + y^2 \geq \frac{1}{2}$ it is clear that $xy \geq \frac{1}{64}$. Applying the oscillatory integral theorem with $k = 3$ we get
$$|F_0| \leq CR^{-\frac{1}{3}}|x-y|^{-\frac{3}{5}\alpha-\frac{9}{10}}.$$
Since $\alpha > \frac{1}{6}$ this gives the right estimate. ∎

We now consider the integral

(5.3.26) $$F = \int_{\beta/2}^{\rho\beta^{3/5}} g_R(t) e^{iR\psi(t)} dt.$$

The oscillatory integral theorem directly applied to this integral fails to give the right estimate. So, we rewrite the phase function in the following way. We have

(5.3.27) $$\psi(t) = t + 2\beta^2 \operatorname{cosec} 2t - \tfrac{1}{2}a^2 \tan t$$

The functions $\tan t$ and $\operatorname{cosec} 2t$ can be expanded in powers of t.

(5.3.28) $$\operatorname{cosec} 2t = \tfrac{1}{2}t^{-1} + \tfrac{1}{3}t + u(t),$$

(5.3.29) $$\tan t = t + v(t)$$

where $u(t) = O(t^3)$, $v(t) = O(t^3)$ for $0 \leq t \leq \frac{\pi}{4}$. Then we have $\psi(t) = \psi_1(t) + \psi_2(t)$ with

(5.3.30) $$\psi_1(t) = \left(1 - \tfrac{1}{2}a^2 + \tfrac{2}{3}\beta^2\right)t + \beta^2 t^{-1},$$

(5.3.31) $$\psi_2(t) = 2\beta^2 u(t) - \tfrac{1}{2}a^2 v(t).$$

Since $|x-y| \leq \frac{1}{2}$ and $a^2 \leq 4$ there are universal constants C_0 and C_1 such that

(5.3.32) $$|\psi_2'(t)| \leq C_0 t^2, \quad |\psi_2'''(t)| \leq C_1 t.$$

We now choose $\delta > 0$ so that $C_0 \delta^{2/5} \leq \frac{4}{9}$ and $C_1 \delta^{2/5} \leq 1$.

Let us write $\nu = (1 - \tfrac{1}{2}a^2 + \tfrac{2}{3}\beta^2)$ so that

(5.3.33) $$F = \int_{\beta/2}^{\rho\beta^{3/5}} \omega_R(t) t^{-\alpha-\frac{3}{2}} e^{iR(\nu t + \beta^2 t^{-1})} e^{iR\psi_2(t)} dt.$$

Making a change of variable we rewrite F as

(5.3.34) $$F = R^{-\alpha-\frac{1}{2}} \beta^{-2\alpha-1} G$$

where G is the integral

(5.3.35) $$G = \int_E \omega(R\beta^2 t) t^{-\alpha-\frac{3}{2}} e^{i(R^2\beta^2 \nu t + \frac{1}{t})} e^{iR\psi_2(R\beta^2 t)} dt$$

the interval E being $\tfrac{1}{2}R^{-1}\beta^{-1} \leq t \leq \rho R^{-1}\beta^{-\frac{7}{5}}$. As the integral taken from one to infinity is bounded we can assume that $R^{-1}\beta^{-\frac{7}{5}} \leq 1$. We now prove

Lemma 5.3.7. *The estimate*

$$|F| \leq C R^{-\frac{1}{3}} |x - y|^{-\alpha - 5/6}$$

is true under the assumption that $\tfrac{1}{2} \leq x^2 + y^2 \leq 4$ *and* $|x - y| \leq 2\delta$.

Proof: First assume that $\nu = (1 - \tfrac{1}{2}a^2 + \tfrac{2}{3}\beta^2)$ is negative. In this case we claim that G is bounded. To see this we calculate

$$-\psi'(t)t^2 = 1 - R^2\beta^2 \nu t^2 + R^2 \beta^2 t^2 \psi_2'(R\beta^2 t).$$

Since $t \leq R^{-1}\beta^{-\frac{7}{5}}$ and $|\psi_2'(t)| \leq C_0 t^2$ we have the bound

(5.3.36) $$R^2\beta^2 t^2 |\psi_2'(R\beta^2 t)| \leq C_0 R^4 \beta^6 (R^{-1}\beta^{-\frac{7}{5}})^4 \leq C_0 \beta^{\frac{2}{5}}.$$

As $\beta \leq \delta$ and $C_0 \delta^{2/5} \leq \tfrac{4}{9}$ we get the bound $|R^2\beta^2 t^2 \psi_2'(R\beta^2 t)| \leq \tfrac{4}{9}$. Therefore, as ν is negative

(5.3.37) $$-\psi'(t) t^2 \geq 1 - \tfrac{4}{9} = \tfrac{5}{9}.$$

We can also prove that $t^3 \psi''(t)$ is bounded. Hence an integration by parts will prove that G is bounded.

Next assume that ν is positive. Now there are two cases to consider.

Case (i). $\nu > 4\beta^{4/5}$. In this case we see that the function $\psi_1(t) = R^2\beta^2\nu t + t^{-1}$ has a stationary point at $t_0 = (R\beta\nu^{\frac{1}{2}})^{-1}$. Since $\nu \geq 4\beta^{4/5}$ we see that $2t_0 \leq R^{-1}\beta^{-7/5}$. We write $G = G_1 + G_2 + G_3$ where they are taken over the interval $\frac{1}{2}R^{-1}\beta^{-1} \leq t \leq \frac{1}{2}t_0$, $\frac{1}{2}t_0 \leq t \leq 2t_0$ and $2t_0 \leq t \leq R^{-1}\beta^{-7/5}$ respectively.

For the first interval we see that

(5.3.38) $$R^2\beta^2\nu t^2 \leq R^2\beta^2\nu \cdot \tfrac{1}{4}(R^2\beta^2\nu)^{-1} = \tfrac{1}{4},$$

(5.3.39) $$R^2\beta^2 t^2 |\psi_2'(R\beta^2 t)| \leq C_0 R^4\beta^6 \cdot \tfrac{1}{16}(R\beta\nu^{\frac{1}{2}})^{-4} \leq \tfrac{1}{16}C_0\beta^2\nu^{-2}.$$

Since $\nu \geq 4\beta^{4/5}$ the above gives

(5.3.40) $$R^2\beta^2 t^2 |\psi_2'(R\beta^2 t)| \leq \frac{C_0}{256}\beta^{\frac{2}{5}} < \tfrac{1}{4}$$

as $\beta \leq \delta_0$. Therefore, the formula

$$-\psi'(t)t^2 = 1 - R^2\beta^2\nu t^2 - R^2\beta^2 t^2 \psi_2'(R\beta^2 t)$$

gives the lower bound $|\psi'(t)t^2| \geq \tfrac{1}{2}$. We also have

(5.3.41) $$t^2\psi''(t) = 2 + R^3\beta^4 t^3 \psi_2''(R\beta^2 t)$$

which gives the bound $|t^3\psi''(t)| \leq C$. Using these bounds for ψ' and ψ'' an integration by parts will prove that G_1 is bounded. Similarly we can prove that G_3 is bounded.

As far as G_2 is concerned we look at the second derivative: for $\frac{1}{2}t_0 \leq t \leq 2t_0$

(5.3.42) $$R^3\beta^4 |\psi_2''(R\beta^2 t)| \leq 2C_1 R^4\beta^6 t_0$$

and since $R\beta = t_0^{-1}\nu^{-1/2}$, and $\nu \geq 4\beta^{4/5}$ we have

(5.3.43) $$R^3\beta^4 |\psi_2''(R\beta^2 t)| \leq 2C_1 t_0^{-3}\nu^{-2}\beta^2 \leq \tfrac{1}{8}C_1 t_0^{-3}\beta^{2/5}.$$

By the choice of δ we have the estimate

(5.3.44) $$|R^3\beta^4 \psi_2''(R\beta^2 t)| \leq \tfrac{1}{8}t_0^{-3}.$$

On the other hand, as $\frac{1}{2}t_0 \leq t \leq 2t_0$

(5.3.45) $$|\psi''(t)| \geq 2t^3 - |R^3\beta^4 \psi_2''(R\beta^2 t)| \geq \tfrac{1}{4}t_0^{-3} - \tfrac{1}{8}t_0^{-3} = \tfrac{1}{8}t_0^{-3}.$$

Therefore, the oscillatory integral theorem gives the estimate

(5.3.46) $$|G_2| \leq Ct_0^{\frac{3}{2}} \int_{\frac{1}{2}t_0}^{2t_0} t^{-\alpha-5/2} dt \leq Ct_0^{-\alpha}.$$

Thus, we have

$$R^{-\alpha-\frac{1}{2}}\beta^{-2\alpha-1}|G_2| \leq CR^{-\alpha-\frac{1}{2}}\beta^{-2\alpha-1}(R\beta\sqrt{\nu})^\alpha \leq CR^{-\frac{1}{2}}|x-y|^{-\alpha-5/6}$$

since ν is bounded by 1.

Case (ii). It still remains to consider the case when $\nu \leq 4\beta^{4/5}$. In this case we take $\rho = \frac{1}{3}$ so that the integral G is defined over the interval $\frac{1}{2}\beta^{-1}R^{-1} \leq t \leq \frac{1}{3}R^{-1}\beta^{-\frac{7}{5}}$. Since now $R^{-1}\beta^{-1}\nu^{-\frac{1}{2}} \geq \frac{1}{2}R^{-1}\beta^{-\frac{7}{5}}$ there is no stationary point for the function $\psi_1(t)$. As $t \leq \frac{1}{3}R^{-1}\beta^{-\frac{7}{5}}$ and $\nu \leq 4\beta^{\frac{4}{5}}$ we see that the term $R^2\beta^2\nu t^2 \leq \frac{4}{9}$ and also the term $R^2\beta^2 t^2 |\psi_2'(R\beta^2 t)| \leq \frac{4}{9}$ by the choice of δ. Hence, $-\psi'(t)t^2 \geq \frac{1}{9}$; we can also get an upper bound for $|t^3 \psi''(t)|$ and now an integration by parts proves that G is bounded. ∎

This completes the proof of Lemma 5.3.6. And all the lemmas put together prove the estimate for J. Thus, Theorem 2.3.1 is completely proved.

5.4 Almost Everywhere and Mean Summability Results

We are now in a position to reap the fruits of our hard work in the previous section. Regarding mean convergence of the Cesàro means $s_N^\alpha f$ we have the following theorem.

Theorem 5.4.1. *Assume that $\alpha > \frac{1}{6}$. Then s_N^α are uniformly bounded on $L^p(\mathbb{R})$ for $1 \leq p \leq \infty$. If $1 \leq p < \infty$ and $f \in L^p(\mathbb{R})$, then $s_N^\alpha f$ converges to f in the norm as $N \to \infty$.*

Proof: The uniform boundedness of the Cesàro means s_N^α follows immediately from the estimates of Theorem 5.3.1. The norm convergence follows by a density argument once we have the following lemma.

Lemma 5.4.1. *For f in $C_0^\infty(\mathbb{R})$ the partial sums $S_N f$ (and hence $s_N^\alpha f$, $\alpha \geq 0$) converge to f in the norm for $1 \leq p < \infty$.*

Proof: Recall that h_k are eigenfunctions of the Hermite operator H. We thus have $H^m h_k = (2k+1)^m h_k$ and consequently

(5.4.1) $$\hat{f}(k) = (2k+1)^{-m}(f, H^m h_k) = (2k+1)^{-m}(H^m f, h_k).$$

As $H^m f$ is in $C_0^\infty(\mathbb{R})$ we get, for any m

(5.4.2) $$|\widehat{f}(k)| \leq (2k+1)^{-m}\|H^m f\|_2.$$

This shows that $S_N f$ converges to f uniformly. The estimate

(5.4.3) $$|S_N f(x)| \leq \Big(\sum_{k=0}^{N}(2k+1)^{-m}|h_k(x)|\Big)\|H^m f\|_2$$

gives, by an application of Holder's inequality for sums, the following:

(5.4.4) $$|S_N f(x)| \leq \|H^m f\|_2 \Big(\sum_{k=0}^{\infty}(2k+1)^{-m}\Big)^{\frac{1}{p'}}$$
$$\times \Big(\sum_{k=0}^{\infty}(2k+1)^{-m}|h_k(x)|^p\Big)^{\frac{1}{p}}.$$

If m is large enough

(5.4.5) $$|S_N f(x)|^p \leq C \sum_{k=0}^{\infty}(2k+1)^{-m}|h_k(x)|^p$$

and the right hand side will be an integrable function. Now dominated convergence theorem proves that $S_N f$ converges to f in the norm. ∎

In order to obtain almost everywhere convergence results we look at the maximal operator associated to the Cesàro means. We define

(5.4.6) $$s_*^\alpha f(x) = \sup_{N \geq 0} |s_N^\alpha f(x)|.$$

Let Mf stand for the Hardy–Littlewood maximal function defined by

(5.4.7) $$Mf(x) = \sup_{r>0} \frac{1}{2r}\int_{-r}^{r}|f(x-y)|dy.$$

We can now prove the following theorem.

Theorem 5.4.2. *Assume that $\alpha > \frac{1}{6}$. Then the maximal operator s_*^α is bounded on $L^p(\mathbb{R})$ for $p > 1$ and is weak type $(1,1)$, i.e.,*

$$\big|\{s_*^\alpha f(x) > \lambda\}\big| \leq C\frac{\|f\|_1}{\lambda}$$

for f in $L^1(\mathbb{R})$. Consequently, $s_N^\alpha f(x)$ converges to $f(x)$ almost everywhere for $1 \leq p \leq \infty$.

Proof: Again by the estimate of Theorem 5.3.1 and standard arguments it follows that

(5.4.8) $$s_*^\alpha f(x) \leq C(Mf(x) + Mf(-x)).$$

This proves that s_*^α is bounded on $L^p(\mathbb{R})$, $p > 1$ and s_*^α is weak type $(1,1)$. Since $s_N^\alpha f(x)$ converges to $f(x)$ uniformly for $C_0^\infty(\mathbb{R})$ functions, it follows, again by a density argument, that $s_N^\alpha f(x)$ converges to $f(x)$ almost everywhere. ∎

Next we proceed to prove the analogues of the Fejer–Lebesgue theorem and the Riemann localization principle. In the case of Fourier series (on the one dimensional torus), the Fejer–Lebesgue theorem states that the Riesz means $S_N^\alpha f(x)$ with $\alpha > 0$ converges to $f(x)$ whenever x is a Lebesgue point of f, for f in L^p, $1 \leq p \leq \infty$. Here, we say that x is a Lebesgue point of f if

(5.4.9) $$\lim_{r \to 0} \frac{1}{r} \int_{|x-y|<r} |f(x) - f(y)|\, dy = 0.$$

The Riemann localization principle states that $S_N f(x)$ converges to 0 if f vanishes in a neighborhood of x. The following is the analogue of these results for the Hermite series.

Theorem 5.4.3. *Assume that $\alpha > \frac{1}{6}$ and f is in $L^p(\mathbb{R})$, $1 \leq p \leq \infty$. If x and $-x$ are both Lebesgue points of f then $s_N^\alpha f(x)$ converges to $f(x)$ as $N \to \infty$; if f vanishes near x and $-x$ then $s_N^\alpha f(x)$ converges to 0.*

Observe that unlike the case of Fourier series we have to assume both x and $-x$ are Lebesgue points of f in order to get the convergence of $s_N^\alpha f(x)$ to $f(x)$; a similar remark applies to the localization result. This extra condition near $-x$ is explained by the appearance of the term $N^{\frac{1}{2}}(1 + N^{\frac{1}{2}}|x+y|)^{-\alpha-5/6}$ in the estimate of the kernel $s_N^\alpha(x,y)$. To establish the above theorem we need the following proposition. We define

(5.4.10) $$G(x) = \sup_{0 < r < 1} \left(\frac{1}{r} \int_{|x-y|<r} |f(y)|\, dy \right).$$

Proposition 5.4.1. *Assume that $\alpha > \frac{1}{6}$ and $f \in L^p(\mathbb{R})$, $1 \leq p \leq \infty$. Then there is a $C > 0$ such that*

(5.4.11) $$|s_N^\alpha f(x)| \leq C(\|f\|_p + G(x) + G(-x)).$$

Proof: For $k \geq 1$ we define

$$f_k(y) = f(y), \quad \text{if } 2^{-k} \leq |x - y| \leq 2^{-k+1}$$
$$= 0, \quad \text{otherwise.}$$

We also set $f_0(y) = f(y) - \sum_{k=1}^{\infty} f_k(y)$. Now

$$(5.4.12) \quad N^{\frac{1}{2}} \int_{-\infty}^{\infty} (1 + N^{\frac{1}{2}}|x - y|)^{-\alpha - 5/6} |f(y)| dy$$
$$\leq N^{\frac{1}{2}} \int_{-\infty}^{\infty} (1 + N^{\frac{1}{2}}|x - y|)^{-\alpha - 5/6} |f_0(y)| dy$$
$$+ \sum_{k=1}^{\infty} N^{\frac{1}{2}} \int_{-\infty}^{\infty} (1 + N^{\frac{1}{2}}|x - y|)^{-\alpha - 5/6} |f_k(y)| dy.$$

The first term is bounded by

$$N^{\frac{1}{2}} \left(\int_{|x-y| \geq 1} (N^{\frac{1}{2}}|x-y|)^{-\alpha p' - 5/6 p'} dy \right)^{\frac{1}{p'}} \|f\|_p$$

which is bounded by some constant times $\|f\|_p$, as the integral is convergent since $\alpha > \frac{1}{6}$.

By the definition of $G(x)$ it follows that $\|f_k\|_1 \leq CG(x) 2^{-k}$ and hence

$$(5.4.13) \quad \sum_{k=1}^{\infty} N^{\frac{1}{2}} \int_{-\infty}^{\infty} (1 + N^{\frac{1}{2}}|x - y|)^{-\alpha - 5/6} |f_k(y)| dy$$
$$\leq CG(x) \sum_{k=1}^{\infty} N^{\frac{1}{2}} 2^{-k} (1 + N^{\frac{1}{2}} 2^{-k})^{-\alpha - 5/6}.$$

The sum on the right hand side converges even when it is extended from $-\infty$ to ∞, provided $\alpha > \frac{1}{6}$. It is then a bounded function of N as it is locally bounded and remains unchanged if N is replaced by $2^{2k} N$. This proves that

$$(5.4.14) \quad N^{\frac{1}{2}} \int_{-\infty}^{\infty} (1 + N^{\frac{1}{2}}|x - y|)^{-\alpha - 5/6} |f(y)| dy \leq C(\|f\|_p + G(x)).$$

A similar argument proves that

$$(5.4.15) \quad N^{\frac{1}{2}} \int_{-\infty}^{\infty} (1 + N^{\frac{1}{2}}|x + y|)^{-\alpha - 5/6} |f(y)| dy \leq C(\|f\|_p + G(-x)).$$

In view of Theorem 5.3.1 this proves the proposition. ∎

We are ready to prove Theorem 5.4.3. The theorem is clearly true for C_0^∞ functions. Replacing $f(y)$ by $f(y) - f(x)g(y) - f(-x)g(-y)$ where g is a C_0^∞ function such that $g(x) = 1$ and $g(-x) = 0$ we can assume that $f(x) = f(-x) = 0$. Thus we are given that

(i) $\lim_{r \to 0} \dfrac{1}{r} \displaystyle\int_{|x-y|<r} |f(y)|dy = 0,$ and

(ii) $\lim_{r \to 0} \dfrac{1}{r} \displaystyle\int_{|x+y|<r} |f(y)|dy = 0.$

Let B be the Banach space of measurable functions with the norm

(5.4.16) $\qquad \|f\|_B = \|f\|_p + G(x) + G(-x)$

where $G(x)$ is defined by (5.4.10). Any f satisfying (i) and (ii) above clearly belongs to the closure of $B \cap C_0^\infty$ in B.

Now $s_N^\alpha f(x)$ can be thought of as a sequence of linear functionals on B and by the above proposition we have

(5.4.17) $\qquad |s_N^\alpha f(x)| \leq C\|f\|_B,$

which means that the sequence is uniformly bounded. Since $s_N^\alpha g(x) \to g(x)$ for g in $C_0^\infty \cap B$ and our f belongs to the closure of $C_0^\infty \cap B$ the theorem follows. ∎

We conclude this chapter with the following result concerning the convergence of $s_N^\alpha f$ when α is below the critical index.

Theorem 5.4.4. *Assume that* $0 < \alpha < \frac{1}{6}$. *The Cesàro means* s_N^α *are uniformly bounded on* $L^p(\mathbb{R})$ *for p in the interval* $\frac{4}{3+6\alpha} < p < \frac{4}{1-6\alpha}$.

This theorem follows by analytic interpolation. For $\alpha = 0$ we know by Theorem 3.1.1 that s_N^α are uniformly bounded on L^p for $4/3 < p < 4$ and for $\alpha > \frac{1}{6}$ they are uniformly bounded on all L^p, $1 \leq p \leq \infty$. An interpolation argument applied to s_N^α will now prove the theorem.

CHAPTER 6
LAGUERRE EXPANSIONS

In this chapter we study three types of Laguerre expansions. The first two types are related to the special Hermite expansions whereas the third one is related to the Hermite expansions. We introduce a convolution structure, called the Laguerre convolution, to study the first type of expansions. The second type of expansions are called the standard Laguerre expansions. We assume a transplantation theorem to be proved in the next chapter and using that we deduce results for standard Laguerre expansions from the results on special Hermite expansions. The third type of expansions are also studied using a transplantation theorem; the results on them will be deduced from the theory of Hermite expansions.

6.1 The Laguerre Convolution

In the study of special Hermite expansions the twisted convolution has played an important role. In the same way the Laguerre convolution which we are going to define shortly is essential for the study of Laguerre expansions of certain type. To motivate the definition of the Laguerre convolution let us consider the twisted convolution $f \times g$ when f and g are radial functions:

$$(6.1.1) \qquad f \times g(z) = \int_{\mathbb{C}^n} f(z-w)g(w)e^{i/2 Imz\cdot\overline{w}}\, dw.$$

It is clear that $f \times g$ is a radial function and changing into polar coordinates we can write the above in the form

$$(6.1.2) \quad f \times g(r) = (2\pi)^{n-\frac{1}{2}} \int_0^\infty \int_0^\pi f\left((r^2 + s^2 - 2rs\cos\theta)^{\frac{1}{2}}\right)$$
$$\cdot j_{n-3/2}(\tfrac{1}{2}rs\sin\theta)g(s)(\sin\theta)^{2n-2}s^{2n-1}d\theta ds.$$

where we have written $j_\alpha(t) = J_\alpha(t)t^{-\alpha}$ and $r = |z|$, $s = |w|$. The above form of the twisted convolution suggests the following definition.

For $\alpha \geq 0$ we define the Laguerre translation $T_x^\alpha f(y)$ of a function f defined on $R_+ = [0,\infty)$ by

$$(6.1.3) \qquad T_x^\alpha f(y) = \frac{\Gamma(\alpha+1)2^\alpha}{\sqrt{2\pi}} \int_0^\pi f\left((x^2 + y^2 + 2xy\cos\theta)^{\frac{1}{2}}\right)$$
$$\cdot j_{\alpha-\frac{1}{2}}(xy\sin\theta)\sin^{2\alpha}\theta d\theta.$$

If f and g are functions defined on R_+ the Laguerre convolution $f \times g$ is defined by

$$(6.1.4) \qquad f \times g(x) = \int_0^\infty T_x^\alpha f(y) g(y) y^{2\alpha+1} dy.$$

Clearly, this is a generalization of the twisted convolution and so we use the same notation to denote both. The above definition makes sense for a suitable class of functions.

The Laguerre convolution operator was introduced by McCully [Mc] for $\alpha = 0$ and extended by Askey [A] to $\alpha > -\frac{1}{2}$. The natural setting to study these operators is the space $L^p(\mu)$ where $d\mu = y^{2\alpha+1} dy$ on $\mathbb{R}_+$. We follow Gorlich-Markett [GM1] in the study of these operators. In (6.1.3) the definition of the Laguerre convolution can be extended to $\alpha > -\frac{1}{2}$. It can even be extended to the critical case $\alpha = -\frac{1}{2}$ but the proofs of the following lemma and theorem are valid only for $\alpha \geq 0$.

We first obtain a representation of the Laguerre translation as an integral operator. Substituting $z^2 = x^2 + y^2 + 2xy \cos \theta$ in (6.1.3) so that for $x, y > 0$, we have the relations $xy \cos \theta = (z^2 - x^2 - y^2)/2$, and

$$(6.1.5) \quad xy \sin \theta = \tfrac{1}{2}\{2(x^2 y^2 + y^2 z^2 + z^2 x^2) - x^4 - y^4 - z^4\}^{\frac{1}{2}} = \rho(x,y,z)$$

it follows that

$$(6.1.6) \qquad T_x^\alpha f(y) = \frac{\Gamma(\alpha+1) 2^\alpha}{\sqrt{2\pi}(xy)^{2\alpha}} \int_{|x-y|}^{(x+y)} f(z) J_{\alpha-\frac{1}{2}}(\rho(x,y,z))$$
$$\times (\rho(x,y,z))^{\alpha-\frac{1}{2}} z \, dz.$$

Thus the Laguerre translation can be represented by

$$(6.1.7) \qquad T_x^\alpha f(y) = \int_0^\infty f(z) K(x,y,z) z^{2\alpha+1} dz,$$

where the kernel K is defined by

$$(6.1.8) \qquad K(x,y,z) = \frac{\Gamma(\alpha+1) 2^\alpha}{\sqrt{2\pi}(xyz)^{2\alpha}} J_{\alpha-\frac{1}{2}}(\rho) \rho^{\alpha-\frac{1}{2}}$$

for $|x - y| \leq z \leq x + y$ and $K(x,y,z) = 0$ otherwise. We now have the following estimate for the kernel K.

Lemma 6.1.1. *Let $\alpha \geq 0$. One has for all $x, y > 0$*

$$\int_0^\infty |K(x,y,z)| z^{2\alpha+1} dz \leq 1.$$

Proof: We use the estimate

$$|J_{\alpha-\frac{1}{2}}(y)| \leq \frac{2^{-\alpha+\frac{1}{2}}}{\Gamma(\alpha+\frac{1}{2})} y^{\alpha-\frac{1}{2}}, \quad y > 0$$

which is valid for $\alpha \geq 0$. This gives us

(6.1.9)
$$\int_0^\infty |K(x,y,z)| z^{2\alpha+1} dz$$
$$\leq \frac{\Gamma(\alpha+1)}{\sqrt{\pi}\Gamma(\alpha+\frac{1}{2})} (xy)^{-2\alpha} \int_0^\infty (\rho(x,y,z))^{2\alpha-1} z\, dz.$$

Going back to the θ variable and using $z\,dz = xy \sin\theta\, d\theta$ we have

(6.1.10)
$$\int_0^\infty |K(x,y,z)| z^{2\alpha+1} dz \leq \frac{\Gamma(\alpha+1)}{\Gamma(\frac{1}{2})\Gamma(\alpha+\frac{1}{2})} \int_0^{2\pi} \sin^{2\alpha}\theta\, d\theta = 1$$

since

$$\int_0^{2\pi} \sin^{2\alpha}\theta\, d\theta = B(\alpha+\tfrac{1}{2}, \tfrac{1}{2}) = \frac{\Gamma(\alpha+\frac{1}{2})\Gamma(\frac{1}{2})}{\Gamma(\alpha+1)}.$$

This proves the lemma. ∎

Let us denote the norm in $L^p(\mathbb{R}_+, d\mu)$ by $\|f\|_{p,\mu}$. Thus $\|f\|_{p,\mu}^p = \int_0^\infty |f(x)|^p x^{2\alpha+1} dx$. With this notation we have the following theorem.

Theorem 6.1.1. *Let $\alpha \geq 0$ and $1 \leq p \leq \infty$. Then*

(i) $\|T_x^\alpha f\|_{p,\mu} \leq \|f\|_{p,\mu}$,

(ii) $\|f \times g\|_{p,\mu} \leq \|g\|_{1,\mu} \|f\|_{p,\mu}$.

Proof: To prove (i) we just use the previous Lemma. In fact, when $p = 1$

$$\|T_x^\alpha f\|_{1,\mu} \leq \int_0^\infty \int_0^\infty |f(z)|\, |K(x,y,z)| z^{2\alpha+1} y^{2\alpha+1} dy\, dz$$

and by the symmetry of the kernel

$$\int_0^\infty |K(x,y,z)|y^{2\alpha+1}dy = \int_0^\infty |K(x,z,y)|y^{2\alpha+1}dy \leq 1$$

by the Lemma. Thus it follows that $\|T_x^\alpha f\|_{1,\mu} \leq \|f\|_{1,\mu}$. When $p = \infty$ it is evident that

$$|T_x^\alpha f(y)| \leq \|f\|_{\infty,\mu} \int_0^\infty |K(x,y,z)|z^{2\alpha+1}dz \leq \|f\|_{\infty,\mu}.$$

Interpolating these two estimates we get (i). To prove (ii) we use Minkowski's integral inequality:

$$\|f \times g\|_{p,\mu} = \left\| \int_0^\infty T_x^\alpha f(y)g(y)y^{2\alpha+1}dy \right\|_p$$

$$\leq \int_0^\infty \|T_x^\alpha f(y)\|_p |g(y)|y^{2\alpha+1}dy$$

$$\leq \|f\|_{p,\mu}\|g\|_{1,\mu}.$$

This completes the proof of the Theorem. ∎

For a moment let us return to the special Hermite expansion of a function f on $\mathbb{C}^n$ in the form

(6.1.11) $$f(z) = (2\pi)^{-n} \sum_{k=0}^\infty f \times \varphi_k(z).$$

We claim that when f is a radial function the above reduces to a Laguerre expansion, a prototype of the expansions which we intend to study in the next section. To prove the claim recall from Chapter 1 that the functions $\mathcal{L}_k^\alpha$ form an orthonormal basis for $L^2(\mathbb{R}_+, dx)$. Therefore, if we let

(6.1.12) $$\psi_k(r) = \left(\frac{2^{1-n}k!}{(k+n-1)!}\right)^{\frac{1}{2}} L_k^{n-1}(\tfrac{1}{2}r^2)e^{-\frac{1}{4}r^2}$$

then it follows that the collection $\{\psi_k(r)\}$ forms an orthonormal basis for $L^2(\mathbb{R}_+, r^{2n-1}dr)$.

If now f is a radial function on $\mathbb{C}^n$ we can expand $f(r) = f(|z|)$ in terms of ψ_k. Thus we have the expansion

(6.1.13) $$f(r) = \sum_{k=0}^\infty \left(\int_0^\infty f(s)\psi_k(s)s^{2n-1}ds\right)\psi_k(r).$$

If we define $R_k(f)$ by the formula

$$(6.1.14) \qquad R_k(f) = \frac{2^{1-n}k!}{(k+n-1)} \int_0^\infty f(s) L_k^{n-1}(\tfrac{1}{2}s^2) e^{-\frac{1}{4}s^2} s^{2n-1} ds$$

then we can write (6.1.13) in the form

$$(6.1.15) \qquad f(z) = \sum_{k=0}^\infty R_k(f)\varphi_k(z).$$

In view of the relation $\varphi_k \times \varphi_j = (2\pi)^n \delta_{jk} \varphi_k$ which follows from $W(\varphi_k) = (2\pi)^n P_k$, (see Theorem 1.3.6) we obtain

$$(6.1.16) \qquad f \times \varphi_k(z) = (2\pi)^n R_k(f)\varphi_k(z).$$

This proves that the special Hermite expansion (6.1.11) reduces to the Laguerre expansion (6.1.15). Hence the claim.

The above consideration suggests that we study expansions in terms of the functions ψ_k^α defined as follows.

$$(6.1.17) \qquad \psi_k^\alpha(x) = \left(\frac{\Gamma(k+1)}{\Gamma(k+\alpha+1)}\right)^{\frac{1}{2}} L_k^\alpha(x^2) e^{-\frac{1}{2}x^2}, \quad \text{where } \alpha > -1.$$

These functions form an orthonormal basis for $L^2(\mathbb{R}_+, x^{2\alpha+1} dx)$ and when $\alpha = n-1$ they are related to $\psi_k(r)$ defined in (6.1.12). In the next section we study the almost everywhere and mean convergence of expansions in terms of ψ_k^α. The following formula is crucial in the study of the above mentioned expansions. To state the formula let us define another related family of functions $\widetilde{\psi}_k^\alpha$,

$$(6.1.18) \qquad \widetilde{\psi}_k^\alpha(x) = \frac{\Gamma(k+1)\Gamma(\alpha+1)}{\Gamma(k+\alpha+1)} L_k^\alpha(x^2) e^{-\frac{1}{2}x^2}.$$

For these functions we have the following theorem.

Theorem 6.1.2. Let $\alpha > -\frac{1}{2}$. Then one has

$$T_x^\alpha \widetilde{\psi}_k^\alpha(y) = \widetilde{\psi}_k^\alpha(x) \widetilde{\psi}_k^\alpha(y).$$

Proof: We first prove this when $\alpha = n-1$, $n = 1, 2, \ldots$ in which case the proof is based on the identity $\varphi_j \times \varphi_k = \delta_{jk}(2\pi)^n \varphi_k$. We consider the radial function

$$(6.1.19) \qquad F_k(s) = \int_{|w'|=1} \varphi_k(z-w) e^{i\frac{1}{2}s \,\mathrm{Im}\, z \cdot w'} dw'$$

where $w = sw'$ and $|z| = r$. Expanding F_k in terms of the functions ψ_j we see that

(6.1.20) $$\int_0^\infty F_k(s)\psi_j(s)s^{2n-1}ds = \varphi_k \times \varphi_j(z) = (2\pi)^n \delta_{jk}\varphi_k(z).$$

Thus it follows that

(6.1.21) $$F_k(s) = (2\pi)^n \varphi_k(z)\psi_k(s).$$

Performing the w' integration in (6.1.19), recalling the definition of the Laguerre convolution and dividing by appropriate constant on both sides of (6.1.21) we get the theorem when $\alpha = n - 1$.

For the general case we need the following Proposition known as the product formula for the Laguerre polynomials. This was proved in Watson [W1].

Proposition 6.1.1. *(Hardy-Watson) Assume $\alpha > -\frac{1}{2}$. Then*

$$\frac{\Gamma(k+1)}{\Gamma(k+\alpha+1)} L_k^\alpha(x^2) L_k^\alpha(y^2)$$
$$= \frac{2^\alpha}{\sqrt{2\pi}} \int_0^\pi L_k^\alpha(x^2 + y^2 + 2xy\cos\theta) e^{-xy\cos\theta}$$
$$\times j_{\alpha-\frac{1}{2}}(xy\sin\theta) \sin^{2\alpha}\theta \, d\theta.$$

Proof: The proof is based on the generating function identity

(6.1.22) $$\sum_{k=0}^\infty \frac{\Gamma(k+1)}{\Gamma(k+\alpha+1)} L_k^\alpha(x^2) L_k^\alpha(y^2) t^k$$
$$= (1-t)^{-1}(ixy\sqrt{t})^{-\alpha} e^{\frac{t}{t-1}(x^2+y^2)} J_\alpha\left(\frac{2ixy\sqrt{t}}{1-t}\right)$$

and the formula

(6.1.23) $$J_\alpha\left((z^2+w^2)^{\frac{1}{2}}\right)$$
$$= \frac{(z^2+w^2)^{\alpha/2}}{\sqrt{2\pi}} \int_0^\pi e^{iz\cos\theta} j_{\alpha-\frac{1}{2}}(w\sin\theta) \sin^{2\alpha}\theta \, d\theta.$$

If we take $z = \frac{1+t}{1-t} ixy$ and $w = xy$ in the above formula we obtain

(6.1.24) $$J_\alpha\left(\frac{2ixy\sqrt{t}}{1-t}\right) = \frac{2^\alpha}{\sqrt{2\pi}}(1-t)^{-\alpha}(ixy\sqrt{t})^\alpha$$
$$\times \int_0^\pi e^{-\frac{1+t}{1-t}xy\cos\theta} j_{\alpha-\frac{1}{2}}(xy\sin\theta) \sin^{2\alpha}\theta \, d\theta.$$

From this (6.1.22) takes the form

$$(6.1.25) \quad \sum_{k=0}^{\infty} \frac{\Gamma(k+1)}{\Gamma(k+\alpha+1)} L_k^\alpha(x^2) L_k^\alpha(y^2) t^k$$

$$= \frac{2^\alpha}{\sqrt{2\pi}}(1-t)^{-\alpha-1} \int_0^\pi e^{\frac{t}{t-1}(x^2+y^2+2xy\cos\theta)} e^{-xy\cos\theta}$$

$$\times j_{\alpha-\frac{1}{2}}(xy\sin\theta) \sin^{2\alpha}\theta\, d\theta.$$

Finally, using the formula

$$(6.1.26) \quad (1-t)^{-\alpha-1} e^{\frac{t}{t-1}x} = \sum_{k=0}^{\infty} L_k^\alpha(x) t^k$$

in (6.1.25) and equating the coefficients on both sides we obtain the proposition. ∎

We can now complete the proof of Theorem 6.1.2 in the general case. Taking $f(x) = L_k^\alpha(x^2)e^{-\frac{1}{2}x^2}$ in (6.1.3) we have

$$(6.1.27) \quad T_x^\alpha f(y) = \frac{\Gamma(\alpha+1)2^\alpha}{\sqrt{2\pi}} \cdot e^{-\frac{1}{2}(x^2+y^2)}$$

$$\times \int_0^\pi L_k^\alpha(x^2+y^2+2xy\cos\theta) e^{-xy\cos\theta}$$

$$\times j_{\alpha-\frac{1}{2}}(xy\sin\theta)\sin^{2\alpha}\theta\, d\theta.$$

In view of the proposition this becomes

$$(6.1.28) \quad T_x^\alpha f(y) = \frac{\Gamma(\alpha+1)\Gamma(k+1)}{\Gamma(k+\alpha+1)} L_k^\alpha(x^2) e^{-\frac{1}{2}x^2} \cdot L_k^\alpha(y^2) e^{-\frac{1}{2}y^2}.$$

If we divide both sides by $\dfrac{\Gamma(k+\alpha+1)}{\Gamma(k+1)\Gamma(\alpha+1)}$ we obtain

$$T_x^\alpha \widetilde{\psi}_k^\alpha(y) = \widetilde{\psi}_k^\alpha(x)\widetilde{\psi}_k^\alpha(y). \quad \blacksquare$$

6.2 Laguerre Expansions of Convolution Type

In this section we consider Cesàro summability of expansions in terms of the functions ψ_k^α. Given $f \in L^p(\mathbb{R}_+, x^{2\alpha+1}dx)$ we have the formal expansion

$$(6.2.1) \quad f(x) = \sum_{k=0}^{\infty} (f, \psi_k^\alpha)\psi_k^\alpha(x)$$

where (f, ψ_k^α) is the inner product in $L^2(\mathbb{R}_+, x^{2\alpha+1} dx)$. In the usual way we define the Cesàro means

$$(6.2.2) \qquad \sigma_N^\delta f(x) = \frac{1}{A_N^\delta} \sum_{k=0}^N A_{N-k}^\delta (f, \psi_k^\alpha) \psi_k^\alpha(x).$$

Regarding the norm convergence of the Cesàro means we have the following result.

Theorem 6.2.1. Let $\alpha \geq 0$ and $\delta > \alpha + \frac{1}{2}$. Then σ_N^δ are uniformly bounded on $L^p(\mathbb{R}_+, x^{2\alpha+1} dx)$ for $1 \leq p \leq \infty$. If $p < \infty$, $\sigma_N^\delta f$ converges to f in the norm as $N \to \infty$.

Proof: Expressing ψ_k^α in terms of $\tilde{\psi}_k^\alpha$ we see that

$$(6.2.3) \qquad \sigma_N^\delta f(x) = \frac{1}{A_N^\delta} \sum_{k=0}^N A_{N-k}^\delta \frac{\Gamma(k+\alpha+1)}{\Gamma(k+1)(\Gamma(\alpha+1))^2} (f, \tilde{\psi}_k^\alpha) \tilde{\psi}_k^\alpha(x).$$

which shows that

$$(6.2.4) \qquad \sigma_N^\delta f(x) = \int_0^\infty K_N^\delta(x, y) f(y) y^{2\alpha+1} dy$$

where the kernel is defined by

$$(6.2.5) \qquad K_N^\delta(x, y) = \frac{1}{A_N^\delta} \sum_{k=0}^N A_{N-k}^\delta \frac{\Gamma(k+\alpha+1)}{\Gamma(k+1)(\Gamma(\alpha+1))^2} \tilde{\psi}_k^\alpha(x) \tilde{\psi}_k^\alpha(y).$$

If we use Theorem 6.1.2 we can write the kernel as

$$(6.2.6) \qquad K_N^\delta(x, y) = \frac{1}{\Gamma(\alpha+1)} T_x^\alpha k_N^\delta(y)$$

where $k_N^\delta(y)$ is defined by

$$k_N^\delta(y) = \frac{1}{A_N^\delta} \sum_{k=0}^N A_{N-k}^\delta L_k^\alpha(y^2) e^{-\frac{1}{2} y^2}$$
$$= \frac{1}{A_N^\delta} L_N^{\alpha+\delta+1}(y^2) e^{-\frac{1}{2} y^2}.$$

Thus $\sigma_N^\delta f$ is given by Laguerre convolution with the kernel k_N^δ:

$$(6.2.7) \qquad \sigma_N^\delta f(x) = \frac{1}{\Gamma(\alpha+1)} f \times k_N^\delta(x).$$

As in the proof of Theorem 2.5.1 we can use the asymptotic properties of the Laguerre polynomial $L_N^{\alpha+\delta+1}$ to show that

$$(6.2.8) \qquad \int_0^\infty |k_N^\delta(y)| y^{2\alpha+1} dy \leq C$$

uniformly in N when $\delta > \alpha + \frac{1}{2}$. Now we can appeal to Theorem 6.1.1 to conclude that σ_N^δ are uniformly bounded on L^p, $1 \leq p \leq \infty$. The norm convergence can be proved using a density argument in a routine way. ∎

Using norm estimates for the Laguerre functions ψ_k^α it can be shown that $\alpha + \frac{1}{2}$ is the critical index for the Cesàro (or Riesz) summability for expansions in terms of ψ_k^α. Concerning the summability below the critical index we can prove the following theorem.

Theorem 6.2.2. *Let $\alpha \geq 0$ and $0 < \delta \leq \alpha + \frac{1}{2}$. The Cesàro means σ_N^δ are uniformly bounded on $L^p(\mathbb{R}_+, x^{2\alpha+1} dx)$ iff p lies in the interval*

$$\frac{4\alpha+4}{2\alpha+3+2\delta} < p < \frac{4\alpha+4}{2\alpha+1-2\delta}.$$

Proof: The necessity of the above condition on p can be proved proceeding as in the case of the Hermite series (Theorem 5.1.2) and using the estimates of Lemma 1.5.4. To prove the sufficiency we use the method of restriction theorems as in the proof of Theorem 2.5.3 for the special Hermite expansions. Again we use Lemma 1.5.4 to prove that

$$(6.2.9) \qquad \|(f, \psi_k^\alpha)\psi_k^\alpha\|_{2,\mu} \leq Ck^{(\alpha+1)(\frac{1}{p}-\frac{1}{2})-\frac{1}{2}} \|f\|_{p,\mu}$$

for $f \in L^p(\mathbb{R}_+, x^{2\alpha+1} dx)$, $1 \leq p < \frac{4\alpha+4}{2\alpha+1}$ where $\|f\|_{p,\mu}$ stands for the norm of f in $L^p(\mathbb{R}_+, x^{2\alpha+1} dx)$. The asymptotic estimates of Lemma 1.5.3 can be used to show that the kernel k_N^δ of the Cesàro means satisfy

$$(6.2.10) \qquad |k_N^\delta(y)| \leq CN^{\alpha+1}(1 + N^{\frac{1}{2}} y)^{-\delta-\alpha-\frac{4}{3}}.$$

The estimates (6.2.9) and (6.2.10) can be used to prove the theorem. The proof is similar to that of Theorem 2.5.3. The details are omitted. ∎

Next we turn our attention to the almost everywhere convergence of the Cesàro means $\sigma_N^\delta f(x)$. In order to study the maximal operator associated to $\sigma_N^\delta f$ we need one more definition, namely, that of generalized Euclidean convolution. Consider now the Euclidean convolution

$$f * g(z) = \int_{\mathbb{C}^n} f(z-w) g(w) dw$$

of two radial integrable functions on $\mathbb{C}^n$, $n \geq 1$. It is then clear that $f * g$ is again radial and is given by a formula like (6.1.2) with the factor $j_{n-3/2}(xy\sin\theta)$ dropped.

By allowing n to be an arbitrary real parameter ≥ 1 we then come to the well known generalized convolution structure

$$(6.2.11) \qquad f * g(x) = \int_0^\infty T_x^E f(y) g(y) y^{2\alpha+1} dy$$

where $T_x^E f(y)$, $x \geq 0$ are the generalized (Euclidean) translation operators

$$(6.2.12) \qquad T_x^E f(y) = \frac{\Gamma(\alpha+1)}{\sqrt{\pi}\Gamma(\alpha+\frac{1}{2})} \int_0^\pi f((x^2+y^2+2xy\cos\theta)^{\frac{1}{2}}) \sin^{2\alpha}\theta d\theta.$$

The range of the parameter α in (6.2.12) can be enlarged to include $\alpha \geq 0$. It can be shown that $L^1(\mathbb{R}_+, x^{2\alpha+1}dx)$ with the above convolution structure (6.2.11) becomes a commutative Banach algebra.

From our point of view the most important connection between the Laguerre convolution and the Euclidean convolution is expressed by the inequality

$$(6.2.13) \qquad |f \times g(x)| \leq |f| * |g|(x)$$

which is implied by $|T_x^\alpha f(y)| \leq T_x^E |f|(y)$ which in turn is a consequence of the estimate for the Bessel function. In view of the above relation, we like to dominate the maximal operator associated to σ_N^δ by another maximal operator defined by the generalized convolution. With this in mind we define the maximal operator

$$(6.2.14) \qquad f^*(x) = \sup_{\varepsilon>0} \frac{1}{\mu(0,\varepsilon)} \int_0^\varepsilon T_x^E(|f|)(y) d\mu(y)$$

where $d\mu$ is the measure $x^{2\alpha+1}dx$. We can also define the dilations $f_\varepsilon(x)$, $\varepsilon > 0$, by setting

$$(6.2.15) \qquad f_\varepsilon(x) = \varepsilon^{-2\alpha-2} f(x/\varepsilon).$$

The following proposition justifies the consideration of this operator.

Proposition 6.2.1. *Suppose φ is a positive non increasing function on $\mathbb{R}_+$ with $\|\varphi\|_{1,\mu} < \infty$. Then*

$$\sup_{\varepsilon>0} |f * \varphi_\varepsilon(x)| \leq \|\varphi\|_{1,\mu} f^*(x).$$

Proof: When φ is a simple function of the form $\sum_{j=1}^n c_j \mathcal{X}_{[0,a_j]}$ where c_j and a_j are positive reals it is straightforward to verify the proposition. The general case can be proved by a limiting argument. ∎

We need to introduce one more maximal function defined on a space of homogeneous type. By this we mean a topological space X equipped with a continuous pseudometric ρ and a positive measure m satisfying

(6.2.16) $$m(B_{2\varepsilon}(x)) \leq Cm(B_\varepsilon(x))$$

with a constant C independent of x and $\varepsilon > 0$. Here

$$B_\varepsilon(x) = \{y \in X : \rho(x,y) < \varepsilon\}.$$

Let (X, ρ, m) be a space of homogeneous type. For any locally integrable function f on X define

(6.2.17) $$Mf(x) = \sup_{\varepsilon>0} m(B_\varepsilon(x))^{-1} \int_{B_\varepsilon(x)} |f(y)| dm(y).$$

It is well known that the maximal function M is weak type $(1,1)$ and is bounded on $L^p(X, dm)$ for $1 < p < \infty$. Here we are concerned with the maximal operator defined by $dm(x) = x^{2\alpha+1} dx$. It is clear that this measure satisfies the doubling condition (6.2.16).

Our goal is to show that the maximal operator associated to $\sigma_N^\delta f$ is majorized by Mf. The following proposition and theorem are taken from Stempak [KS]. We first show that f^* is majorized by Mf. We start with the following proposition.

Proposition 6.2.2. *There exists a constant $C > 0$ independent of $0 < t < x$ such that*

$$\|T_x^E \mathcal{X}_{[0,t]}\|_\infty \leq C(t/x)^{2\alpha+1}.$$

Proof: From the definition it follows easily that

(6.2.18) $$T_{xt}^E \mathcal{X}_{[0,t]}(ty) = T_x^E \mathcal{X}_{[0,1]}(y)$$

and consequently it is enough to prove the proposition when $t = 1$ and so $x > 1$. We can further assume that x is large enough, say $x > 10$.

The function $T_x^E \mathcal{X}_{[0,1]}$ is supported in $[x-1, x+1]$. In terms of the variable z we have for $x - 1 \leq y \leq x + 1$

(6.2.19) $\quad T_x^E \mathcal{X}_{[0,1]}(y) = C(\alpha) x^{-2\alpha-1} \cdot \dfrac{x}{y} \int_{|x-y|}^1 \left(\dfrac{\rho(x,y,z)}{y} \right)^{2\alpha-1} z \, dz.$

Since $\frac{x}{y} < 2$ we need only check that the integrals on the right side of (6.2.19) are uniformly bounded on $x, y > 0$, $|x - y| < 1$. When $\alpha \geq \frac{1}{2}$ this follows from the estimate $\rho(x,y,z) \leq yz$. When $-\frac{1}{2} < \alpha < \frac{1}{2}$ we have

$$4(\rho(x,y,z))^2 = 2(x^2 y^2 + y^2 z^2 + z^2 x^2) - x^4 - y^4 - z^4$$
$$= ((x+y)^2 - z^2)(z^2 - (x-y)^2).$$

Since $|x - y| < z < 1$ we have $(x+y)^2 - z^2 > 4y^2$ and consequently $\rho(x,y,z) \geq y(z^2 - b^2)^{\frac{1}{2}}$ where $b = |x - y|$. Therefore, the integral is bounded by

$$\int_b^1 (z^2 - b^2)^{\alpha - \frac{1}{2}} z \, dz \leq C$$

since $\alpha > -\frac{1}{2}$. This completes the proof. ∎

Theorem 6.2.3. *There exists a constant $C > 0$ such that $f^*(x) \leq CMf(x)$. Consequently, f^* is weak type $(1,1)$ and strong type (p,p), for $1 < p < \infty$.*

Proof: We can assume that $f \geq 0$ Define the auxiliary maximal functions

$$f_1^*(x) = \sup_{\varepsilon < x} \dfrac{1}{\mu(0,\varepsilon)} \int_0^\varepsilon T_x^E f(y) d\mu(y),$$

and

$$f_2^*(x) = \sup_{\varepsilon \geq x} \dfrac{1}{\mu(0,\varepsilon)} \int_0^\varepsilon T_x^E f(y) d\mu(y).$$

We will show that $f_i^*(x) \leq C_i M f(x)$ for $i = 1, 2$. When $\varepsilon \geq x$, $\varepsilon^{-2\alpha-2} \leq 2^{2\alpha+2} (\varepsilon + x)^{-2\alpha-2}$ and $T_x^E \mathcal{X}_{[0,\varepsilon]} \leq \mathcal{X}_{[0,\varepsilon+x]}$. We write

$$\dfrac{1}{\mu(0,\varepsilon)} \int_0^\varepsilon T_x^E f(y) y^{2\alpha+1} dy = \dfrac{2\alpha+2}{\varepsilon^{2\alpha+2}} (T_x^E f, \mathcal{X}_{[0,\varepsilon]})$$
$$= \dfrac{2\alpha+2}{\varepsilon^{2\alpha+2}} (f, T_x^E \mathcal{X}_{[0,\varepsilon]})$$
$$\leq C(\varepsilon + x)^{-2\alpha-2} \int_0^{x+\varepsilon} f(y) y^{2\alpha+1} dy.$$

This proves that $f_2^*(x) \leq C_2 M f(x)$.

When $0 < \varepsilon < x$, $T_x^E \mathcal{X}_{[0,\varepsilon]}$ is supported in $[x - \varepsilon, x + \varepsilon]$. By Proposition 6.2.2

$$(T_x^E f, \mathcal{X}_{[0,\varepsilon]}) = (f, T_x^E \mathcal{X}_{[0,\varepsilon]}) \leq C \left(\frac{\varepsilon}{x}\right)^{2\alpha+1} (f, \mathcal{X}_{[x-\varepsilon, x+\varepsilon]}).$$

Further, $x^{2\alpha+1}\varepsilon \geq C\mu(x - \varepsilon, x + \varepsilon)$ as $\varepsilon < x$. Therefore

$$\frac{1}{\mu(0,\varepsilon)} \int_0^\varepsilon T_x^E f(y) y^{2\alpha+1} dy \leq \frac{C}{x^{2\alpha+2}\varepsilon} \int_{x-\varepsilon}^{x+\varepsilon} f(y) y^{2\alpha+1} dy$$

$$\leq C \mu(B_\varepsilon(x))^{-1} \int_{B_\varepsilon(x)} f(y) d\mu$$

$$\leq C M f(x).$$

This completes the proof. ∎

Finally we come to the almost everywhere convergence of the Cesàro means $\sigma_N^\delta f(x)$.

Theorem 6.2.4. *Assume that $\alpha \geq 0$ and $f \in L^p$.*

(i) *If $\delta > \alpha + \frac{2}{3}$ and $1 \leq p \leq \infty$ then $\sigma_N^\delta f(x)$ converges to $f(x)$ almost everywhere.*

(ii) *If $\delta > \alpha + \frac{1}{2}$ and $p > \frac{4}{3}$ then $\sigma_N^\delta f(x)$ converges to $f(x)$ almost everywhere.*

Proof: It is enough to show that the maximal operator

$$\sigma_*^\delta f(x) = \sup_{N > 0} |\sigma_N^\delta f(x)|$$

is weak type (p,p). First assume that $\delta > \alpha + \frac{2}{3}$. Then in view of the estimate (6.2.10) and the Proposition 6.2.1 it follows that $\sigma_*^\delta f(x) \leq Cf^*(x)$. Theorem 6.2.3 shows that $\sigma_*^\delta f$ is weak type (p,p) for $1 \leq p \leq \infty$.

When $\delta > \alpha + \frac{1}{2}$ we can split the kernel into three parts and proceed as in the proof of Theorem 2.5.2. The first and third parts will be of weak type (p,p), $1 \leq p \leq \infty$. The middle one can be shown to be dominated by $(M|f|^p)^{\frac{1}{p}}$ provided $p > \frac{4}{3}$. The details are omitted. This completes the proof of the theorem. ∎

6.3 Standard Laguerre Expansions

In this section we study expansions in terms of the functions $\mathcal{L}_k^\alpha$, which form an orthonormal basis for $L^2(\mathbb{R}_+, dx)$. We also study the n dimensional case of this type of expansions. As usual we are interested in the norm convergence of the series

(6.3.1) $$f = \sum_{k=0}^{\infty} (f, \mathcal{L}_k^\alpha) \mathcal{L}_k^\alpha$$

for functions in $L^p(\mathbb{R}_+)$, $1 \leq p \leq \infty$. Just like the case of one dimensional Hermite series we have the following theorem concerning the norm convergence of (6.3.1).

Theorem 6.3.1. *(Askey-Wainger) Let $\alpha \geq 0$. Then the series (6.3.1) converge to f in the norm if and only if $4/3 < p < 4$.*

We refer to the paper [AW] for a proof of this result. This result was proved in 1965 and later Muckenhoupt ([Mu1] and [Mu2]) considered weighted norm inequalities for the partial sums of the above series. We are rather interested in the behavior of the Cesàro means associated to the above series. These are defined by

(6.3.2) $$C_N^\delta f = \frac{1}{A_N^\delta} \sum_{k=0}^{N} A_{N-k}^\delta (f, \mathcal{L}_k^\alpha) \mathcal{L}_k^\alpha.$$

In 1972, Poiani [Po] studied the Cesàro means of order $\delta = 1$ and obtained the following weighted inequality.

Theorem 6.3.2. *(Poiani) Let $\alpha > -1$ and β satisfy the conditions $-\frac{2}{3p} - \frac{1}{2} \leq \beta \leq -\frac{2}{3p} + \frac{7}{6}$ and*

$$-\frac{1}{p} - \min\left(\frac{\alpha}{2}, \frac{1}{4}\right) < \beta < 1 - \frac{1}{p} + \min\left(\frac{\alpha}{2}, \frac{1}{4}\right)$$

where $1 \leq p \leq \infty$. Then there is a constant C independent of N such that

$$\int_0^\infty |C_N^1 f(x) x^\beta|^p dx \leq C \int_0^\infty |f(x) x^\beta|^p dx.$$

The proof depends heavily on the asymptotic properties of the Laguerre functions on various intervals on the half line. An exact expression for the Cesàro kernel was obtained by Campbell. This expresison together with estimates for $\mathcal{L}_k^\alpha$ led Poiani to get rather precise estimates for the Cesàro kernel. We are not going to prove the above theorem. But we need the result in the proof of the transplantation theorems for the Laguerre expansions. Cesàro means of fractional orders were studied by Markett in a series of papers [Ma1], [Ma2] and [GM2]. He used the Laguerre convolution structure studied in Section 6.1 to prove the following result.

Theorem 6.3.3. *(Markett) Let $\alpha \geq 0$ and $0 < \delta < \frac{1}{2}$. Then the Cesàro means C_N^δ are uniformly bounded on $L^p(\mathbb{R}_+)$ if and only if p satisfies $\frac{4}{3+2\delta} < p < \frac{4}{1-2\delta}$. When $\delta > \frac{1}{2}$, C_N^δ are uniformly bounded on all $L^p(\mathbb{R}_+)$, $1 \leq p \leq \infty$.*

Markett has also obtained precise estimates for the norms of the Cesàro means on $L^p(\mathbb{R}_+)$ as N tends to infinity. The above theorem shows that $\delta = \frac{1}{2}$ is the critical index for the Cesàro summability of the standard Laguerre expansions. In proving the above theorem we will not follow Markett. Instead, we deduce the theorem from the result of Theorem 6.2.2 using a transplantation theorem which we describe now.

Given $\alpha, \beta > -1$ we define an operator T_α^β formally by setting

$$(6.3.3) \qquad T_\alpha^\beta f = \sum_{k=0}^\infty (f, \mathcal{L}_k^\beta) \mathcal{L}_k^\alpha.$$

This operator was studied by Kanjin [K] where he proved that T_α^β is bounded on $L^p(\mathbb{R}_+)$, $1 < p < \infty$ when both $\alpha, \beta \geq 0$. When $\nu = \min\{\alpha, \beta\}$ satisfies $-1 < \nu < 0$, T_α^β is bounded on $L^p(\mathbb{R}_+)$ for $(1 + \frac{\nu}{2})^{-1} < p < -\frac{2}{\nu}$. We will prove this theorem in the next chapter. Now we will assume this result and see how Theorem 6.3.3 follows from Theorem (6.2.2).

Consider multiplier transforms m_λ^α defined for the Laguerre expansion (6.3.1) by

$$(6.3.4) \qquad m_\lambda^\alpha f = \sum_{k=0}^\infty \lambda(2k+1)(f, \mathcal{L}_k^\alpha) \mathcal{L}_k^\alpha$$

where λ is a bounded function defined on $\mathbb{R}_+$. From the definition of T_α^β it follows that

$$(6.3.5) \qquad T_\beta^\alpha \, m_\lambda^\alpha \, T_\alpha^\beta \, f = m_\lambda^\beta \, f$$

for any $\alpha, \beta > -1$. In view of this relation it is immediate that m_λ^α is bounded on $L^p(\mathbb{R}_+)$ if and only if m_λ^β is bounded on $L^p(\mathbb{R}_+)$. Thus we can transplant any norm inequality for m_λ^α into an inequality for m_λ^β. This justifies the name transplantation operators for T_α^β.

It is clear that C_N^δ are operators of the form m_λ^α and so in order to prove Theorem 6.3.3 it is enough to prove it when $\alpha = 0$. But when $\alpha = 0$, we have the relation $\psi_k^\alpha(x) = \mathcal{L}_k^\alpha(x^2)$ and hence a moment's thought reveals that expansions in terms of ψ_k^α is equivalent to expansions in terms of $\mathcal{L}_k^\alpha$ (when $\alpha = 0$). Therefore, the case $\alpha = 0$ of Theorem 6.3.3 follows from Theorem 6.2.2. Thus if we assume that T_α^β, $\alpha, \beta \geq 0$ are bounded in $L^p(\mathbb{R}_+)$, then Theorem 6.3.3 is completely proved but for the L^1 or L^∞ result which are not accessible through transplantation. ∎

Using the boundedness of the transplantation operators T_α^β on L^p spaces we can also deduce multiplier theorems for standard Laguerre expansions.

Theorem 6.3.4. *Assume $\alpha \geq 0$. Let λ be a bounded, twice differentiable function on $\mathbb{R}_+$ which satisfies (a) $t|\lambda'(t)| \leq C_1$, (b) $t^2|\lambda''(t)| \leq C_2$ for all $t \geq 1$. Then m_λ^α is bounded on $L^p(\mathbb{R}_+)$, $1 < p < \infty$. When $4/3 < p < 4$, m_λ^α is bounded on $L^p(\mathbb{R}_+)$ under the condition (a) alone.*

Proof: Since T_α^β, $\alpha, \beta \geq 0$ are bounded on $L^p(\mathbb{R}_+)$, $1 < p < \infty$, it is enough to show that m_λ^α is bounded on $L^p(\mathbb{R}_+)$ for a particular value of α. When both conditions (a) and (b) are met we show that m_λ^0 is bounded on $L^p(\mathbb{R}_+)$, $1 < p < \infty$; when only (a) is satisfied we show that $m_\lambda^{\frac{1}{2}}$ is bounded on $L^p(\mathbb{R}_+)$, $4/3 < p < 4$.

As we have seen in Section 6.1, the special Hermite expansion reduces to a Laguerre expansion when we consider only radial functions. In view of (6.1.14) and (6.1.16) it follows that when f is radial on $\mathbb{C}$, the special Hermite expansion becomes the expansion in terms of ψ_k^0, which in turn is equivalent to the standard Laguerre expansion (6.3.1) with $\alpha = 0$. Hence the boundedness of m_λ^0 follows from Theorem 2.4.1 by taking f to be a radial function there.

To show that $m_\lambda^{\frac{1}{2}}$ is bounded on $L^p(\mathbb{R}_+)$, for $4/3 < p < 4$ we use Theorem 4.2.1. This theorem states that when λ satisfies condition (a) then the operator

$$(6.3.6) \qquad T_\lambda f = \sum_{k=0}^{\infty} \lambda(k)(f, h_k) h_k$$

satisfies the weighted norm inequality

(6.3.7) $$\int_{-\infty}^{\infty} |T_\lambda f(x)|^p |x|^{-\frac{p}{2}+1} dx \leq C \int_{-\infty}^{\infty} |f(x)|^p |x|^{-\frac{p}{2}+1} dx.$$

Suppose now $f \in L^p(\mathbb{R})$ is an odd function. Then as h_{2k} is even and h_{2k+1} is odd it follows that

(6.3.8) $$T_\lambda f(x) = \sum_{k=0}^{\infty} \lambda(2k+1)(f, h_{2k+1}) h_{2k+1}(x).$$

Using the relation (1.5.53) we calculate that

(6.3.9) $$(f, h_{2k+1}) = (-1)^k (g, \mathcal{L}_k^{\frac{1}{2}})$$

where $g(x) = f(\sqrt{x}) x^{-\frac{1}{4}}$ for $x > 0$. Therefore,

(6.3.10) $$T_\lambda f(\sqrt{x}) x^{-\frac{1}{4}} = \sum_{k=0}^{\infty} \lambda(2k+1)(g, \mathcal{L}_k^{\frac{1}{2}}) \mathcal{L}_k^{\frac{1}{2}}(x)$$

for $x > 0$. In view of this relation the inequality (6.3.7) is equivalent to

(6.3.11) $$\int_0^{\infty} |m_\lambda^{\frac{1}{2}} g(x)|^p dx \leq C \int_0^{\infty} |g(x)|^p dx.$$

This proves that $m_\lambda^{\frac{1}{2}}$ is bounded on $L^p(\mathbb{R}_+)$, when $4/3 < p < 4$.

The above proof demonstrates very well the power of the transplantation technique. Though Theorem 6.3.3 can be proved without recourse to transplantation, it is not so in the case of Theorem 6.3.4. If the reader goes back to the proofs of Theorems 2.4.1 and 4.2.1, he can convince himself that the special values $\alpha = 0$ in Theorem 2.4.1 and $\alpha = \frac{1}{2}$ in Theorem 4.2.1 cannot be dispensed with. Nevertheless, the real power of the transplantation theorem can be seen only when we study the multiple Laguerre expansions. The methods used to study one dimensional Laguerre expansions are no longer useful in the study of multiple Laguerre expansions. The transplantation theorem comes for our rescue and with much ease we can prove theorems for multiple Laguerre expansions. We now proceed to describe these expansions.

Let $\mathbb{R}_+^n = \{x \in \mathbb{R}^n : x_j \geq 0\}$ and consider for every $\alpha \in \mathbb{R}_+^n$ and a multiindex μ the normalized Laguerre functions

(6.3.12) $$\mathcal{L}_\mu^\alpha(x) = \prod_{j=1}^{n} \mathcal{L}_{\mu_j}^{\alpha_j}(x_j), \quad x \in \mathbb{R}_+^n.$$

These functions $\mathcal{L}_\mu^\alpha$ form an orthonormal basis for $L^2(\mathbb{R}_+^n)$ and for f in $L^p(\mathbb{R}_+^n)$ one has the multiple Laguerre expansion

(6.3.13) $$f = \sum_\mu (f, \mathcal{L}_\mu^\alpha) \mathcal{L}_\mu^\alpha.$$

Defining the projections Q_k^α by

(6.3.14) $$Q_k^\alpha f = \sum_{|\mu|=k} (f, \mathcal{L}_\mu^\alpha) \mathcal{L}_\mu^\alpha$$

we can write the expansion (6.3.13) as $f = \sum_{k=0}^\infty Q_k^\alpha f$. We then define the Cesàro means and multipliers in the usual way. Thus

(6.3.15) $$C_N^\delta f = \frac{1}{A_N^\delta} \sum_{k=0}^N A_{N-k}^\delta Q_k^\alpha f,$$

and

(6.3.16) $$m_\lambda^\alpha f = \sum_{k=0}^\infty \lambda(2k+n) Q_k^\alpha f.$$

In order to study these operators we need an n–dimensional version of the transplantation operator T_α^β.

For α and β in $\mathbb{R}_+^n$ we define T_α^β in the most natural way by setting

(6.3.17) $$T_\alpha^\beta f = \sum_\mu (f, \mathcal{L}_\mu^\beta) \mathcal{L}_\mu^\alpha.$$

Then in view of the one dimensional transplantation theorem, it follows by iteration that these operators T_α^β are bounded on $L^p(\mathbb{R}_+^n)$, $1 < p < \infty$. Thus the n-dimensional transplantation theorem comes for free and it is with the help of these operators we can handle the multiple Laguerre expansions. We now state two theorems regarding the expansion (6.3.13).

Theorem 6.3.5. *Let $\alpha \in \mathbb{R}_+^n$ and $\delta > \frac{1}{2}$. The Cesàro means C_N^δ are uniformly bounded on $L^p(\mathbb{R}_+^n)$ if and only if*

$$\frac{4n}{2n+1+2\delta} < p < \frac{4n}{2n-1-2\delta}.$$

In view of the fact that T_α^β is bounded on $L^p(\mathbb{R}_+^n)$ for $\alpha, \beta \in \mathbb{R}_+^n$ and $T_\beta^\alpha m_\lambda^\alpha T_\alpha^\beta f = m_\lambda^\beta f$ it is enough to prove the theorem when $\alpha = (0, 0, \ldots, 0)$. We will show below how that particular case follows from the corresponding result for special Hermite expansions. Using the same arguments and the transplantation theorem we can prove the following multiplier theorem.

Theorem 6.3.6. *Let $\alpha \in \mathbb{R}_+^n$ and $k = n+1$ when n is odd and $k = n+2$ when n is even. Assume that λ is a C^k function on $(0, \infty)$ which satisfies the conditions $\sup_{t>0} |t^j \lambda^{(j)}(t)| \leq C_j$, $j = 0, 1, \ldots, k$. Then m_λ^α is bounded on $L^p(\mathbb{R}_+^n)$, $1 < p < \infty$.*

We will now demonstrate how the above two theorems can be deduced when $\alpha = m$ is a multiindex. To this end we make the following definition. We say that a function f defined on $\mathbb{C}^n$ is m-homogeneous when it satisfies

(6.3.18) $$f(e^{i\theta} z) = e^{im \cdot \theta} f(z)$$

where $e^{i\theta} z = (e^{i\theta_1} z_1, \ldots, e^{i\theta_n} z_n)$. Here $m = (m_1, \ldots, m_n)$ and m_j are integers (positive or negative). It is clear that when f is m-homogeneous it is of the form

(6.3.19) $$f(z) = e^{im \cdot \theta} f_0(r_1, \ldots, r_n), \quad z_j = r_j e^{i\theta_j}.$$

From Theorem 1.3.5 it follows that the special Hermite function $\Phi_{\mu\nu}$ is $(\nu - \mu)$- homogeneous.

Now let us consider the special Hermite expansion of a function f which is m- homogeneous where now m is a multiindex. Then the coefficient $(f, \Phi_{\mu\nu})$ is non zero only when $\nu = \mu + m$. This can be seen by integrating with respect to the θ variable. Hence the special Hermite expansion of an m-homogeneous function takes the form

(6.3.20) $$f(z) = \sum_\mu (f, \Phi_{\mu,\mu+m}) \Phi_{\mu,\mu+m}(z).$$

By formula (ii) of Theorem 1.3.5 it is clear that the above is the expansion of the function g on $\mathbb{R}_+^n$ which is related to f by

$$f_0(r_1, r_2, \ldots, r_n) = g(\tfrac{1}{2} r_1^2, \ldots, \tfrac{1}{2} r_n^2)$$

in terms of $\mathcal{L}_\mu^m(r_1, \ldots, r_n)$. Therefore, it is immediate that Theorem 6.3.5 with $\alpha = m$ follows from Corollary 2.5.1 by considering m-homogeneous functions. ∎

In the same way Theorem 6.3.6 with $\alpha = m$ follows from Theorem 2.4.1. In this particular case, a weaker form of Theorem 6.3.6 is known and is due to Dlugosz [Dl]. She deduced it from a correpsonding theorem on the Heisenberg group. This multiplier theorem with $\alpha = m$

is needed in the proof of the transplantation theorem. In the next section we will prove a weighted version of the above multiplier theorem, again for $\alpha = m$, a multiindex which will be used to prove a weighted transplantation theorem.

6.4 Laguerre Expansions of Hermite Type

In the preceding sections we studied two types of Laguerre expansions and observed that both of them are related to the special Hermite expansions. They arise as particular cases of the special Hermite expansions and this explains why the critical index for Cesàro summability is $(n-\frac{1}{2})$ in the case of multiple Laguerre expansions on $\mathbb{R}^n_+$. On the other hand we have seen in Chapter 3 that the critical index for multiple Hermite expansions is $(\frac{n-1}{2})$ and that for the one dimensional Hermite expansions is $\frac{1}{6}$ which we proved in Chapter 5. Since the Hermite functions h_{2k} and h_{2k+1} satisfy the relations

(6.4.1) $$h_{2k}(x) = (-1)^k 2^{-\frac{1}{2}} \mathcal{L}_k^{-\frac{1}{2}}(x^2)(2x)^{\frac{1}{2}},$$

and

(6.4.1′) $$h_{2k+1}(x) = (-1)^k 2^{-\frac{1}{2}} \mathcal{L}_k^{\frac{1}{2}}(x^2)(2x)^{\frac{1}{2}},$$

we might consider expansions in terms of the functions $\mathcal{L}_k^\alpha(x^2)(2x)^{\frac{1}{2}}$ rather than expansions in terms of $\mathcal{L}_k^\alpha(x)$. The later expansions may have better properties similar to the Hermite case.

This was precisely the point of view taken by Markett [Ma2]. Following Markett let us define the Laguerre functions φ_k^α by

(6.4.2) $$\varphi_k^\alpha(x) = \mathcal{L}_k^\alpha(x^2)(2x)^{\frac{1}{2}}$$

which form an orthonormal basis for $L^2(\mathbb{R}_+)$. For functions f in $L^p(\mathbb{R}_+)$ we consider the expansion

(6.4.3) $$f = \sum_{k=0}^\infty (f, \varphi_k^\alpha) \varphi_k^\alpha.$$

In [Ma2] Markett observed that the operator norms of the partial sum operators associated to the above series acting on $L^1(\mathbb{R}_+)$ grows like $N^{\frac{1}{6}}$ which is better than the growth $N^{\frac{1}{2}}$ for the norms of the partial sum operators associated to the standard Laguerre expansions. This indicated a smaller critical index for the Cesàro summability of the above series (6.4.3). He actually conjectured that the critical index lies between $\frac{1}{6}$ and $\frac{1}{2}$. This conjecture was proved to be correct in [T4] by showing that the critical index is actually $\frac{1}{6}$.

Let $S_R^\delta f$ stand for the Riesz means defined by

$$(6.4.4) \qquad S_R^\delta f = \sum \left(1 - \frac{2k+1}{R}\right)_+^\delta (f, \varphi_k^\alpha) \varphi_k^\alpha.$$

Then one has the following theorem.

Theorem 6.4.1. *Assume that $\alpha \geq -\frac{1}{2}$. If $\alpha > \frac{1}{6}$ then the Riesz means S_R^δ are uniformly bounded on $L^p(\mathbb{R}_+)$ for $1 \leq p \leq \infty$.*

This theorem was proved in [T4] under the assumption that $\alpha \geq \frac{1}{2}$. This was proved using the same method that was used in the case of one dimensional Hermite expansions. We could express the kernel of S_R^δ as an oscillatory integral and estimate the integral using the method of stationary phase. The method yielded the estimate

$$(6.4.5) \qquad |S_R^\delta(x,y)| \leq C R^{\frac{1}{2}} (1 + R^{\frac{1}{2}} |x-y|)^{-\delta - 5/6}$$

when $\frac{1}{6} < \delta < \frac{1}{2}$. From this estimate Theorem 6.4.1 follows immediately. This also leads to almost everywhere convergence results.

Here we will not follow the above method. Instead, we will deduce the above theorem from the correspnding result for the Hermite series using a transplantation operator. This transplantation operator which we denote by τ_α^β is defined by

$$(6.4.6) \qquad \tau_\alpha^\beta f = \sum_{k=0}^\infty (f, \varphi_k^\beta) \varphi_k^\alpha$$

where $\alpha, \beta \geq -\frac{1}{2}$. In the next Chapter we will show that τ_α^β is bounded on $L^p(\mathbb{R}_+)$, $1 < p < \infty$. Assuming this result for the time being we will now show how we deduce Theorem 6.4.1 from Theorem 5.4.1. As in the case of standard Laguerre expansions, we define multiplier transforms M_λ^α by

$$(6.4.7) \qquad M_\lambda^\alpha f = \sum_{k=0}^\infty \lambda(2k+1)(f, \varphi_k^\alpha) \varphi_k^\alpha$$

where λ is a bounded function on $(0,\infty)$. It then follows that

(6.4.8) $$\tau_\beta^\alpha M_\lambda^\alpha \tau_\alpha^\beta f = M_\lambda^\beta f.$$

In view of this and the boundedness of τ_α^β on $L^p(\mathbb{R}_+)$, Theorem 6.4.1 will be proved once we prove it when $\alpha = \frac{1}{2}$.

Since the Riesz means $S_R^\delta f$ defined in (6.4.4) are multipliers of the form $M_\lambda^\alpha f$ it is enough to exhibit a relation between $M_\lambda^{\frac{1}{2}} f$ and the multiplier $T_\lambda f$ defined by

(6.4.9) $$T_\lambda f = \sum_{k=0}^\infty \lambda(k)(f, h_k) h_k$$

for the Hermite series. As in the previous section we take f to be an odd function in (6.4.9). Then it is immediate that

(6.4.10) $$T_\lambda f(x) = \sum_{k=0}^\infty \lambda(2k+1)(f, h_{2k+1}) h_{2k+1}(x).$$

In view of the relation (6.4.1′) one calculates that $(f, h_{2k+1}) h_{2k+1} = (f, \varphi_k^{\frac{1}{2}}) \varphi_k^{\frac{1}{2}}$. Thus

(6.4.11) $$T_\lambda f(x) = \sum_{k=0}^\infty \lambda(2k+1)(f, \varphi_k^{\frac{1}{2}}) \varphi_k^{\frac{1}{2}}(x) = M_\lambda^{\frac{1}{2}} f(x).$$

So, if T_λ is bounded on $L^p(\mathbb{R})$ then $M_\lambda^{\frac{1}{2}}$ is bounded on $L^p(\mathbb{R}_+)$. Theorem 6.4.1 now follows from Theorem 5.4.1 when $\alpha = \frac{1}{2}$. ∎

Thus we are perfectly right in expecting the expansions (6.4.3) to behave better than the standard Laguerre expansions. We can now try our luck in the higher dimensional case also. Let μ be a multiindex. For each $\alpha = (\alpha_1, \ldots, \alpha_n)$ where $\alpha_j \geq -\frac{1}{2}$ we define n-dimensional Laguerre functions Φ_μ^α of Hermite type by

(6.4.12) $$\Phi_\mu^\alpha(x) = \prod_{j=1}^n \varphi_{\mu_j}^{\alpha_j}(x_j), \quad x \in \mathbb{R}_+^n.$$

The name is justified by the fact that the Hermite functions Φ_μ are of the above form with $\alpha_j = \pm\frac{1}{2}$. Let us define the projections P_k^α by the formula $P_k^\alpha f = \sum_{|\mu|=k}(f, \Phi_\mu^\alpha) \Phi_\mu^\alpha$ so that the expansion of a function in

$L^p(\mathbb{R}_+^n)$ takes the form $f = \sum_{k=0}^{\infty} P_k^\alpha f$. We then define S_R^δ and M_λ^α in the usual way:

(6.4.13) $$S_R^\delta f = \sum_{k=0}^{\infty} \left(1 - \frac{2k+n}{R}\right)_+^\delta P_k^\alpha f,$$

and

(6.4.14) $$M_\lambda^\alpha f = \sum_{k=0}^{\infty} \lambda(2k+n) P_k^\alpha f.$$

Regarding the Riesz means and the multiplier we have the following theorems.

Theorem 6.4.2. *Let $\alpha_j \geq -\frac{1}{2}$, $n \geq 2$ and $\delta > \frac{(n-1)}{2}$. Then S_R^δ are uniformly bounded on $L^p(\mathbb{R}_+^n)$ for all $1 < p < \infty$.*

We deduce this theorem from the corresponding result for the multiple Hermite expansions. With that in mind we consider the n-dimensional version of the transplantation operator (6.4.6) defined by

(6.4.15) $$\tau_\alpha^\beta f = \sum_\mu (f, \Phi_\mu^\beta) \Phi_\mu^\alpha.$$

By iteration it follows that τ_α^β is bounded on $L^p(\mathbb{R}_+^n)$ provided $\alpha_j, \beta_j \geq -\frac{1}{2}$ and $1 < p < \infty$. Since we also have an identity of the form (6.4.8) it is enough to prove Theorem 6.4.2 when $\alpha_j = \frac{1}{2}$ for all j. But this can be deduced as in the one dimensional case. By taking f to be a function on $\mathbb{R}^n$ which is odd in each variable separately it follows that the Riesz means $S_R^\delta f$ defined for the Hermite expansions reduces to the Riesz means defined in (6.4.14) with $\alpha = (\frac{1}{2}, \frac{1}{2}, \ldots, \frac{1}{2})$. So Theorem 6.4.2 follows from Theorem 3.3.2. ∎

In a similar way using the Marcinkiewicz multiplier theorem for the Hermite expansions and the transplantation theorem we can obtain the following theorem.

Theorem 6.4.3. *Let k be an integer bigger than $n/2$ and let λ be a C^k function on $(0, \infty)$. Assume that λ satisfies the estimates*

$$\sup_{t>0} |t^j \lambda^{(j)}(t)| \leq C_j,$$

for $j = 0, 1, \ldots, k$. Then the operators M_λ^α are bounded on $L^p(\mathbb{R}_+^n)$, $1 < p < \infty$ provided $\alpha_j \geq -\frac{1}{2}$ for all j.

The above two theorems indicate that the expansions (6.4.13) in terms of Φ_μ^α are the genuine n–dimensional Laguerre expansions. To put it in a different form we can say that the Hermite and special Hermite expansions are the prototypes of Laguerre expansions on $\mathbb{R}^n$ and $\mathbb{C}^n$ respectively. The rest of the section is devoted to proving a weighted version of the multiplier Theorem 6.3.4. We state and prove the following theorem for m_λ^α here rather than in Section 6.3 as the proof involves a Littlewood–Paley–Stein theory for expansions in terms of φ_k^α. Here is the theorem which we need in Chapter 7 in the proof of the transplantation theorem.

Theorem 6.4.4. *Let $\alpha = n - 1$ be an integer and assume that λ is a C^{n+2} function which satisfies the estimates $\sup_{t>0} |t^j \lambda^{(j)}(t)| \leq C_j$, $j = 0, 1, \ldots, (n+2)$. Then the operator m_λ^α satisfies*

$$\int_0^\infty |m_\lambda^\alpha f(x)|^p x^{\frac{p}{4} - \frac{1}{2}} dx \leq C \int_0^\infty |f(x)|^p x^{\frac{p}{4} - \frac{1}{2}} dx$$

for all f in $L^p(\mathbb{R}_+)$, $1 < p < \infty$.

In order to prove this theorem we first establish a weighted inequality for the multiplier transform T_λ defined for the special Hermite expansions. Recall that T_λ is defined by

(6.4.16) $$T_\lambda f(z) = (2\pi)^{-n} \sum_{k=0}^\infty \lambda(2k + n) f \times \varphi_k(z).$$

Theorem 6.4.4 will be deduced from the following result concerning the multiplier T_λ.

Theorem 6.4.5. *Assume that the function λ satisfies the conditions of Theorem 6.4.4. Then*

$$\int_{\mathbb{C}^n} |T_\lambda F(z)|^p |z|^{(n-\frac{1}{2})p - (2n-1)} dz \leq C \int_{\mathbb{C}^n} |F(z)|^p |z|^{(n-\frac{1}{2})p - (2n-1)} dz$$

for all $F \in L^p(\mathbb{C}^n)$, $1 < p < \infty$, which are radial.

Before proceeding to the proof of the above theorem, let us see how Theorem 6.4.4 can be deduced from the above result. When F is a radial function it follows from (6.1.14) and (6.1.16) that $F \times \varphi_k$ is a radial function and

(6.4.17) $$F \times \varphi_k(r) = \frac{2\pi^n k!}{(k+n-1)!}$$
$$\times \left(\int_0^\infty F(s) L_k^{n-1}(\tfrac{1}{2}s^2) e^{-\tfrac{1}{4}s^2} s^{2n-1} ds \right)$$
$$\cdot L_k^{n-1}\left(\tfrac{1}{2}r^2\right) e^{-\tfrac{1}{4}r^2}$$

which can be written as

(6.4.18) $$F \times \varphi_k(r) = c_n(f, \mathcal{L}_k^{n-1}) \mathcal{L}_k^{n-1}(r^2/2) r^{1-n}$$

where c_n is a constant and F and f are related by

(6.4.19) $$f(r^2/2) = F(r) r^{n-1}.$$

Thus the multiplier $T_\lambda F(z)$ takes the form

(6.4.20) $$T_\lambda F(z) = c_n \sum_{k=0}^\infty \lambda(2k+n)(f, \mathcal{L}_k^{n-1}) \mathcal{L}_k^{n-1}(r^2/2) r^{1-n}$$

i.e., $T_\lambda F(z) = c_n m_\lambda^{n-1} f(r^2/2) r^{1-n}$ where $r = |z|$. In view of this relation Theorem 6.4.5 gives

(6.4.21) $$\int_0^\infty |m_\lambda^{n-1} f(r^2/2)|^p r^{(1-n)p} r^{(n-\tfrac{1}{2})p} dr$$
$$\leq C \int_0^\infty |f(r^2/2)|^p r^{(1-n)p} r^{(n-\tfrac{1}{2})p} dr$$

which after a change of variables becomes

(6.4.22) $$\int_0^\infty |m_\lambda^{n-1} f(r)|^p r^{\tfrac{p}{4}-\tfrac{1}{2}} dr \leq C \int_0^\infty |f(r)|^p r^{\tfrac{p}{4}-\tfrac{1}{2}} dr.$$

This completes the proof of Theorem 6.4.4. ∎

Coming to the proof of Theroem 6.4.4 we recall from Chapter 2 that the unweighted inequality for T_λ was proved using Littlewood–Paley–Stein theory. We established the pointwise estimate

(6.4.23) $$g(T_\lambda F, z) \leq C g_k^*(f, z)$$

where $k = n+1$ and from this and the L^p mapping properties of g and g_k^* we established the L^p boundedness of T_λ. In view of (6.4.23), Theorem 6.4.4 can be proved once we have the following proposition. By slightly abusing the notation we write

$$\|F\|_{p,w}^p = \int_{\mathbb{C}^n} |F(z)|^p |z|^{(n-\frac{1}{2})p-(2n-1)} dz. \tag{6.4.24}$$

Proposition 6.4.1.

(i) There are constants C_1 and C_2 such that for $1 < p < \infty$ one has

$$C_1 \|F\|_{p,w} \leq \|g(F)\|_{p,w} \leq C_2 \|F\|_{p,w}$$

for all radial functions F in $L^p(\mathbb{C}^n)$.

(ii) If $p > 2$ and $k > n$ then for all radial functions F in $L^p(\mathbb{C}^n)$ one also has

$$\|g_k^*(F)\|_{p,w} \leq C \|F\|_{p,w}.$$

Proof: We first claim that (i) implies (ii). To see this let h be a nonnegative function. Since $k > n$ it follows that

$$\int_{\mathbb{C}^n} (g_k^*(F,z))^2 h(z) dz \leq C \int_{\mathbb{C}^n} (g(F,z))^2 M h(z) dz \tag{6.4.25}$$

where Mh is the Hardy–Littlewood maximal function of h. For $p > 2$ let $q = p/2$ and define $h_1(z) = h(z)|z|^{(2n-1)(1-\frac{2}{p})}$. Replacing h by h_1 in (6.4.25) we get

$$\int_{\mathbb{C}^n} (g_k^*(F,z))^2 |z|^{(2n-1)(1-\frac{2}{p})} h(z) dz \tag{6.4.26}$$
$$\leq C \int_{\mathbb{C}^n} (g(F,z))^2 M h_1(z) dz.$$

We now write the right hand side of (6.4.26) as

$$\int_{\mathbb{C}^n} (g(F,z))^2 M h_1(z) dz \tag{6.4.27}$$
$$= \int_{\mathbb{C}^n} (g(F,z))^2 |z|^{(2n-1)(1-\frac{2}{p})} |z|^{-(2n-1)/q'} M h_1(z) dz$$

where q' is the conjugate index of q. By applying Holder's inequality we obtain

(6.4.28)
$$\int_{\mathbb{C}^n} (g(F,z))^2 Mh_1(z) dz$$
$$\leq \left(\int_{\mathbb{C}^n} (g(F,z))^p |z|^{(n-\frac{1}{2})p-(2n-1)} dz \right)^{2/p}$$
$$\times \left(\int_{\mathbb{C}^n} (Mh_1(z))^{q'} |z|^{-(2n-1)} dz \right)^{1/q'}.$$

Since $q' > 1$, $-2n < -2n+1 < 2n(q'-1)$ so that the function $|z|^{-(2n-1)}$ is in the $A_{q'}$ class of Muckenhoupt and consequently

(6.4.29)
$$\left(\int_{\mathbb{C}^n} (Mh_1(z))^{q'} |z|^{-(2n-1)} dz \right)^{1/q'}$$
$$\leq C \left(\int_{\mathbb{C}^n} (h_1(z))^{q'} |z|^{-(2n-1)} dz \right)^{1/q'} \leq C \|h\|_{q'}.$$

On the other hand, if F radial, the right hand side inequality of part (i) of the proposition gives

(6.4.30)
$$\int_{\mathbb{C}^n} (g(F,z))^p |z|^{(n-\frac{1}{2})p-(2n-1)} dz$$
$$\leq C \int_{\mathbb{C}^n} |F(z)|^p |z|^{(n-\frac{1}{2})p-(2n-1)} dz.$$

Using (6.4.29) and (6.4.30) in (6.4.26) we have the inequality

(6.4.31) $$\int_{\mathbb{C}^n} (g_k^*(F,z))^2 |z|^{(2n-1)(1-\frac{2}{p})} h(z) dz \leq C \|F\|_{p,w}^2 \|h\|_{q'}.$$

Now taking supremum over all h with $\|h\|_{q'} \leq 1$ we obtain $\|g_k^*(F)\|_{p,w} \leq C \|F\|_{p,w}$ for radial F.

In order to establish part (i) we first observe that $g(F)$ is radial whenever F is radial. In fact an easy calculation shows that

(6.4.32) $$F \times \varphi_k(z) = c_n(h, \varphi_k^{n-1}) \varphi_k^{n-1}(r/\sqrt{2}) r^{-n+\frac{1}{2}}$$

where c_n is a constant and F and h are related by $F(\sqrt{2}r) r^{n-\frac{1}{2}} = h(r)$. Therefore, if $T^t = e^{-tL}$,

(6.4.33) $$T^t F(z) = c_n \widetilde{T}^t h(r/\sqrt{2}) r^{-n+\frac{1}{2}}$$

where $\widetilde{T}^t$ is the semigroup defined by

$$(6.4.34) \qquad \widetilde{T}^t f(r) = \sum_{k=0}^{\infty} e^{-(2k+n)t} (f, \varphi_k^{n-1}) \varphi_k^{n-1}(r).$$

This gives the relation

$$(6.4.35) \qquad g(F, z) = c_n \widetilde{g}(h, r/\sqrt{2}) r^{-n+\frac{1}{2}}$$

where $\widetilde{g}$ is the g function associated to the semigroup $\widetilde{T}^t$. From the above relation it follows that

$$(6.4.36) \qquad \int_{\mathbb{C}^n} (g(F, z))^p |z|^{(n-\frac{1}{2})p - (2n-1)} dz = c_n' \int_0^{\infty} (\widetilde{g}(h, r))^p dr$$

and also we have the equality

$$(6.4.37) \qquad \int_{\mathbb{C}^n} |F(z)|^p |z|^{(n-\frac{1}{2})p - (2n-1)} dz = c_n'' \int_0^{\infty} |h(r)|^p dr.$$

Therefore, part (i) of the proposition is equivalent to the following result.

Proposition 6.4.2. *There are constants C_1 and C_2 such that $C_1 \|h\|_p \leq \|\widetilde{g}(h)\|_p \leq C_2 \|h\|_p$ for h in $L^p(\mathbb{R}_+)$ and $1 < p < \infty$.*

Proof: The L^2 estimate $\|\widetilde{g}(h)\|_2 = 2^{-1} \|h\|_2$ follows from the orthonormality of the functions φ_k^{n-1} and the Plancherel theorem. We only need to show that $\|\widetilde{g}(h)\|_p \leq C_2 \|h\|_p$ as the reverse inequality can be obtained in a routine way. The proof of this is very similar to the proof of the corresponding inequality for the g function defined for the Hermite semigroup.

We consider $\widetilde{g}$ as a singular integral operator whose kernel $K_t(r, s)$ takes values in the Hilbert space $L^2(\mathbb{R}_+, t dt)$. The kernel is given by

$$(6.4.38) \qquad K_t(r, s) = \frac{\partial}{\partial t} \Big\{ \sum_{k=0}^{\infty} e^{-(2k+n)t} \varphi_k^{n-1}(r) \varphi_k^{n-1}(s) \Big\}.$$

Using the generating function (1.1.47) we see that

$$(6.4.39) \qquad \sum_{k=0}^{\infty} \varphi_k^{n-1}(r) \varphi_k^{n-1}(s) w^{2k}$$

$$= 2(rs)^{\frac{1}{2}} w^{-n+1} (1 - w^2)^{-1} e^{-\frac{1}{2} \frac{1+w^2}{1-w^2}(r^2+s^2)}$$

$$\times e^{-i(n-1)\pi/2} J_{n-1}\Big(\frac{2irsw}{1-w^2} \Big).$$

By taking $w = e^{-t}$ in this formula and differentiating with respect to t we obtain an expression for the kernel $K_t(r,s)$. If we set $I_k(z) = e^{-ik\pi/2}J_k(iz)$ then I_k satisfies the following relations and estimate:

(6.4.40) $$2I'_k(z) = I_{k-1}(z) + I_{k+1}(z), \quad k \geq 1$$

(6.4.40') $$I'_0(z) = I_1(z),$$

(6.4.40'') $$|I_k(z)| \leq Cz^{-\frac{1}{2}}e^z, \quad z \geq 1.$$

Using these relations and the estimate we can prove the following:

(6.4.41) $$|K_t(r,s)| \leq Ct^{-\frac{3}{2}}e^{-\frac{a}{t}|r-s|^2},$$

(6.4.41') $$|\partial_r K_t(r,s)| \leq Ct^{-2}e^{-\frac{a}{t}|r-s|^2}.$$

The proof of these estimates are similar to the proof of Lemma 4.2.1. From the above estimates it follows that

(6.4.42) $$\int_0^\infty |K_t(r,s)|^2 t dt \leq C|r-s|^{-2},$$

and

(6.4.42') $$\int_0^\infty |\partial_r K_t(r,s)|^2 t dt \leq C|r-s|^{-4}.$$

Hence K_t is a Calderon–Zygmund kernel and hence $\widetilde{g}$ is bounded on $L^p(\mathbb{R}_+)$. ∎

This completes the proof of Theorem 6.4.4.

CHAPTER 7
THE TRANSPLANTATION THEOREMS

The aim of this chapter is to prove the two transplantation theorems we have used in the previous chapter. The proofs of these two theorems are lengthy and involve many identities and estimates satisfied by Laguerre and Bessel functions. We also need the multiplier Theorems 6.3.6 and 6.4.4 when α is a nonnegative integer. The basic idea behind the proofs is simple; only the lengthy identities and estimates make it look complicated.

7.1 The Transplantation Operators

In the previous chapter we introduced the transplantation operator T_α^β and used its L^p boundedness in deducing results for multiple Laguerre expansions. This operator T_α^β was studied by Kanjin [K] and here we reproduce his proof of the L^p boundedness of T_α^β. Recall from (6.3.2) that T_α^β is defined formally by

$$(7.1.1) \qquad T_\alpha^\beta f = \sum_{k=0}^\infty (f, \mathcal{L}_k^\beta) \mathcal{L}_k^\alpha.$$

Here $\alpha, \beta > -1$ and $(f, \mathcal{L}_k^\beta)$ is the Laguerre coefficient

$$(7.1.2) \qquad (f, \mathcal{L}_k^\beta) = \int_0^\infty f(x) \mathcal{L}_k^\beta(x) dx.$$

We note that the integral converges and $(f, \mathcal{L}_k^\beta)$ finite when $\beta \geq 0$ and $1 \leq p \leq \infty$ or $-1 < \beta < 0$ and $(1 + \beta/2)^{-1} < p \leq \infty$. Regarding the L^p boundedness of T_α^β we prove the following theorem.

Theorem 7.1.1. (Kanjin) Let $\alpha, \beta > -1$ and $\gamma = \min\{\alpha, \beta\}$. If $\gamma \geq 0$ then T_α^β is bounded on $L^p(\mathbb{R}_+)$ for all $1 < p < \infty$; if $-1 < \gamma < 0$, T_α^β is bounded on $L^p(\mathbb{R}_+)$ provided $(1 + \frac{\gamma}{2})^{-1} < p < -\frac{2}{\gamma}$.

The proof of this theorem is long and will be presented in several steps. Before proceeding to the proof let us consider another transplantation operator τ_α^β which was also introduced in the previous chapter.

This operator τ_α^β was defined for Laguerre expansions of Hermite type. Formally,

(7.1.3) $$\tau_\alpha^\beta f = \sum_{k=0}^\infty (f, \varphi_k^\beta) \varphi_k^\alpha$$

where now α and β are assumed to be bigger than or equal to $-\frac{1}{2}$. The boundedness of τ_α^β on $L^p(\mathbb{R}_+)$ has been already used in the study of Laguerre expansions of Hermite type. Here we prove the following theorem.

Theorem 7.1.2. Assume that $\alpha, \beta \geq -\frac{1}{2}$. Then τ_α^β is bounded on $L^p(\mathbb{R}_+)$, $1 < p < \infty$.

Both transplantation theorems will be proved simultaneously. In fact Theorem 7.1.2 is equivalent to a weighted version of Theorem 7.1.1. This can be seen as follows. Since $\varphi_k^\alpha(x) = \mathcal{L}_k^\alpha(x^2)(2x)^{\frac{1}{2}}$ one calculates that

(7.1.4) $$(f, \varphi_k^\alpha) = \frac{1}{\sqrt{2}}(\widetilde{f}, \mathcal{L}_k^\alpha)$$

where $\widetilde{f}(x) = f(\sqrt{x})x^{-\frac{1}{4}}$, $x > 0$. Thus one has the relation

(7.1.5) $$\tau_\alpha^\beta f(\sqrt{x})x^{-\frac{1}{4}} = T_\alpha^\beta \widetilde{f}(x).$$

From this it follows that the inequality

(7.1.6) $$\int_0^\infty |\tau_\alpha^\beta f(x)|^p dx \leq C \int_0^\infty |f(x)|^p dx$$

holds if and only if

(7.1.7) $$\int_0^\infty |T_\alpha^\beta \widetilde{f}(x)|^p x^{\frac{p}{4}-\frac{1}{2}} dx \leq C \int_0^\infty |\widetilde{f}(x)|^p x^{\frac{p}{4}-\frac{1}{2}} dx$$

holds. This fact will be made use of in the proof of Theorem 7.1.2.

We start with the following observation. Since $(T_\alpha^\beta f, g) = (f, T_\beta^\alpha g)$ it is enough to prove Theorem 7.1.1 when $\alpha < \beta$. We claim that in order to prove Theorem 7.1.1 it is enough to consider the operators T_n^β and $T_\alpha^{\alpha+2}$ where n is a nonnegative integer, $n \leq \beta \leq n+2$ and $-1 < \alpha < 0$. To see this we proceed in the following way.

Case (i). When $0 \leq \alpha < \beta$ we find integers n and m so that $2n \leq \alpha < 2n+2$, $2m \leq \beta < 2m+2$. We can write T_α^β as

(7.1.8) $$T_\alpha^\beta = T_\alpha^{2n} \cdot T_{2n}^{2n+2} \cdots T_{2m-2}^{2m} \cdot T_{2m}^\beta.$$

Case (ii). When $-1 < \alpha < 0 \leq \beta$, let $2m \leq \beta < 2m+2$. Then we can write

(7.1.9) $$T_\alpha^\beta = T_\alpha^{\alpha+2} \cdot T_{\alpha+2}^0 \cdot T_0^2 \cdots T_{2m}^\beta.$$

Case (iii). When $-1 < \alpha < \beta < 0$ we can write

(7.1.10) $$T_\alpha^\beta = T_\alpha^{\alpha+2} \cdot T_{\alpha+2}^0 \cdot T_0^{\beta+2} \cdot T_{\beta+2}^\beta.$$

Thus in view of (7.1.8), (7.1.9) and (7.1.10) it follows that we only need to consider the operators T_n^β, $n \leq \beta \leq n+2$ and $T_\alpha^{\alpha+2}$, $-1 < \alpha < 0$. A similar remark applies to the operators τ_α^β.

With this initial reduction we proceed to study the operators T_n^β, $T_\alpha^{\alpha+2}$, τ_n^β and $\tau_\alpha^{\alpha+2}$. These operators will be studied using analytic interpolation. As such the first step in the proof is to extend the definition of T_α^β and τ_α^β to complex values of β. By the explicit formula

(7.1.11) $$L_n^\beta(x) = \sum_{k=0}^n \binom{n+\beta}{n-k} \frac{(-x)^k}{k!}$$

the definition of the Laguerre polynomial is extended to complex β. For fixed $x > 0$, $L_n^\beta(x)$ is analytic in β except at the points $\beta = -n-1$, $-n-2, \ldots$. The coefficient

$$\left(\frac{\Gamma(n+1)}{\Gamma(n+\beta+1)}\right)^{\frac{1}{2}}$$

is analytic in β in the cut plane $|\arg(\beta+n+1)| < \pi$, where we take the branch of the square root equal to $+1$ when $\beta = 0$. Therefore, if f is in $C_0^\infty(\mathbb{R}_+)$ and $\beta = \alpha + \delta + i\theta$, θ real, for the coefficients $(f, \mathcal{L}_n^\beta)$ and (f, φ_n^β) are well defined. In the next section we show that for each $j = 1, 2, \ldots$ one has the estimates

(7.1.12) $$|(f, \mathcal{L}_n^\beta)| \leq C(1+|\theta|)^j e^{\frac{\pi}{2}|\theta|} n^{\frac{\delta-j}{2}+\frac{1}{4}}$$

valid for large n. In view of this estimate it follows that $T_\alpha^\beta f$ and $\tau_\alpha^\beta f$ are in $L^2(\mathbb{R}_+)$ when $f \in C_0^\infty(\mathbb{R}_+)$ even if β is complex.

The main step in the study of the operators T_α^β and τ_α^β is to prove the L^p mapping properties of the operators $T_\alpha^{\alpha+k+i\theta}$ and $\tau_\alpha^{\alpha+k+i\theta}$ for $k = 0, 2$. In what follows $M(\theta)$ will stand for a generic function of θ which may vary from place to place but is of admissible growth, i.e., it satisfies the estimate

(7.1.13) $$\log M(2\theta) e^{-a|\theta|} \leq C$$

for all θ real with $a < \pi$. Regarding the operators $T_\alpha^{\alpha+k+i\theta}$ and $\tau_\alpha^{\alpha+k+i\theta}$ we prove the following two propositions.

Proposition 7.1.1. *Let* $\alpha = 0, 1, 2, \ldots, f \in C_0^\infty(\mathbb{R}_+)$ *and* $1 < p < \infty$. *Then one has for* $k = 0, 2$

(i) $\|T_\alpha^{\alpha+k+i\theta} f\|_p \leq M(\theta) \|f\|_p$,

(ii) $\|\tau_\alpha^{\alpha+k+i\theta} f\|_p \leq M(\theta) \|f\|_p$.

Proposition 7.1.2.

(i) *If* $\alpha \geq 0$, $f \in C_0^\infty(\mathbb{R}_+)$ *and* $1 < p < \infty$ *then* $\|T_\alpha^{\alpha+2} f\|_p \leq C \|f\|_p$; *if* $-1 < \alpha < 0$ *then the above inequality is valid for* $\left(1 + \frac{\alpha}{2}\right)^{-1} < p < -\frac{2}{\alpha}$.

(ii) *If* $\alpha \geq -\frac{1}{2}$, $f \in C_0^\infty(\mathbb{R}_+)$ *and* $1 < p < \infty$ *then* $\|\tau_\alpha^{\alpha+2} f\|_p \leq C \|f\|_p$.

These two propositions will be proved in the following sections. Here we assume them and show how the transplantation theorem can be deduced from them. The operators $T_\alpha^{\alpha+2}$ and $\tau_\alpha^{\alpha+2}$ are bounded on the appropriate L^p spaces in view of Proposition 7.1.2. Therefore, in view of the initial reductions we have made we only need to prove that T_n^β and τ_n^β with $n \leq \beta \leq n+2$ are bounded on $L^p(\mathbb{R}_+)$, $1 < p < \infty$. To this end we use analytic interpolation.

For $f, g \in C_0^\infty(\mathbb{R}_+)$ and $z = \delta + i\theta$ complex we consider the function

(7.1.14) $$\Phi_\alpha(z) = \int_0^\infty T_\alpha^{\alpha+2z} f(x) g(x) dx.$$

Then it follows from the estimate (7.1.12) with $j = 4$ that

(7.1.15) $$|\Phi_\alpha(z)|^2 = \Big|\sum_{n=0}^\infty (f, \mathcal{L}_n^{\alpha+2z})(g, \mathcal{L}_n^\alpha)\Big|^2$$
$$\leq \Big(\sum_{n=0}^\infty |(f, \mathcal{L}_n^{\alpha+2z})|^2\Big) \Big(\sum_{n=0}^\infty |(g, \mathcal{L}_n^\alpha)|^2\Big)$$
$$\leq C(1+\theta^4)^2 e^{2\pi|\theta|} \|g\|_2^2$$

for $0 \leq \delta \leq 1$ and $\theta \in \mathbb{R}$ where C is a constant not depending on δ and θ. Thus, $\Phi_\alpha(z)$ is analytic in the strip $0 < \delta < 1$, continuous in the closed strip and of admissible growth there, that is,

$$\sup\left\{e^{-a|\theta|}\log|\Phi_\alpha(z)| : 0 \leq \delta \leq 1, -\infty < \theta < \infty\right\}$$

is finite with $a < \pi$. Let $\alpha = n$ be a nonnegative integer, $1 < p < \infty$ and $1/p + 1/q = 1$. By Proposition 7.1.1 we have the estimate

(7.1.16) $\qquad |\Phi_\alpha(k + i\theta)| \leq M(2\theta), \quad k = 0, 1$

for $\|f\|_p \leq 1$, $\|g\|_q \leq 1$. By the analytic interpolation theorem of Stein it follows that $\|\Phi_\alpha(\delta)\| \leq C$ for $0 \leq \delta \leq 1$ where C is a constant depending on δ. Thus we have

(7.1.17) $\qquad \|T_n^\beta f\|_p \leq C\|f\|_p, \quad f \in C_0^\infty(\mathbb{R}_+)$

for β satisfying $n \leq \beta \leq n + 2$ and $1 < p < \infty$.

This completes the proof of Theorem 7.1.1 modulo the Propositions 7.1.1 and 7.1.2. A similar argument shows that τ_n^β, $n \leq \beta \leq n + 2$ is bounded on $L^p(\mathbb{R}_+)$, $1 < p < \infty$ and that will prove Theorem 7.1.2. The propositions will be proved in several steps in the following sections.

7.2 Further Reduction in the Proof of the Propositions

In this section we reduce the propositions 7.1.1 and 7.1.2 to estimating a related operator $\widetilde{T}_\alpha^\beta$. To effect this reduction we make use of the multiplier operators m_λ^α and M_λ^α. We start with two lemmas. The first Lemma proves the estimate (7.1.12) which is necessary to extend T_α^β to complex values of β.

Lemma 7.2.1. Let $f \in C_0^\infty(\mathbb{R}_+)$, $\alpha > -1$ and $\beta = \alpha + \delta + i\theta$. Then for every $j = 1, 2, \ldots$ there are a constant C and number n_0 such that

$$|(f, \mathcal{L}_n^\beta)| \leq C(1 + |\theta|)^j e^{\frac{\pi}{2}|\theta|} n^{\frac{\delta-j}{2} + \frac{1}{4}}$$

for $n \geq n_0$ and $\theta \in \mathbb{R}$.

Proof: The proof needs the formula (see [EM1])

(7.2.1) $\qquad L_n^\beta(y)e^{-y}y^\beta = \frac{(n-j)!}{n!}\left(\frac{d}{dy}\right)^j\left\{L_{n-j}^{\beta+j}(y)e^{-y}y^{\beta+j}\right\}.$

Using this we have

$$(7.2.2) \quad (f, \mathcal{L}_n^\beta) = \left(\frac{\Gamma(n+1)}{\Gamma(n+\beta+1)}\right)^{\frac{1}{2}} \frac{(n-j)!}{n!}(-1)^j$$
$$\times \int_0^\infty \left(\frac{d}{dy}\right)^j (f(y)y^{-\beta/2}e^{\frac{y}{2}})L_{n-j}^{\beta+j}(y)e^{-y}y^{\beta+j}dy.$$

Since f is a function with compact support in $(0, \infty)$ we may assume that supp $f \subset [a, b]$, $0 < a < b < \infty$. Thus

$$(7.2.3) \quad |(f, \mathcal{L}_n^\beta)| \leq C(1+|\theta|)^j n^{-j} \left|\left(\frac{\Gamma(n+1)}{\Gamma(n+\beta+1)}\right)^{\frac{1}{2}}\right| \int_a^b |L_{n-j}^{\beta+j}(y)|dy$$

where C is a constant independent of n and θ. We now apply the formula (see [EM1])

$$(7.2.4) \quad L_m^{\mu+\nu}(y) = \frac{\Gamma(m+\mu+\nu+1)}{\Gamma(\nu)\Gamma(m+\nu+1)} \int_0^1 v^\mu (1-v)^{\nu-1} L_m^\nu(vy)dv,$$

Re $\mu > -1$, Re $\nu > -1$ with $\mu = \alpha + j - 1$, $\nu = 1 + \delta + i\theta$ and $m = n - j$ to the integrand $L_{n-j}^{\beta+j}(y)$. Then

$$(7.2.5) \quad |(f, \mathcal{L}_n^\beta)| \leq \frac{C(1+|\theta|)^j}{|\Gamma(1+\delta+i\theta)|} n^{-j} A_{n,\beta}$$
$$\times \int_a^b \int_0^1 v^{\alpha+j-1} |L_{n-j}^{\alpha+j-1}(vy)|dv$$

where

$$A_{n,\beta} = \frac{|\Gamma(n+\beta+1)|}{|\Gamma(n+\alpha)|} \left|\left(\frac{\Gamma(n+1)}{\Gamma(n+\beta+1)}\right)^{\frac{1}{2}}\right|.$$

Expressing $(\Gamma(1+\delta+i\theta))^{-1}$ in terms of the beta function and using $|\Gamma(\frac{1}{2}+i\theta)|^2 = \pi \operatorname{sech}\pi\theta$ we have

$$(7.2.6) \quad |\Gamma(1+\delta+i\theta)|^{-1} \leq Ae^{\frac{\pi}{2}|\theta|}$$

where A is an absolute constant.

If we use Stirling's formula for the gamma functions to estimate $A_{n,\beta}$ it is not difficult to see that

$$(7.2.7) \quad A_{n,\beta} \leq Cn^{(\delta-\alpha+2)/2}$$

THE TRANSPLANTATION THEOREMS 173

for $n \geq n_0$ where the constant C and n_0 depends only on δ and α. To estimate the integral on the right side of (7.2.5) we note that the integral is independent of β. We have

$$\int_a^b \int_0^1 v^{\alpha+j-1}|L_{n-j}^{\alpha+j-1}(vy)|dvdy = \int_0^1 \Big(\int_{av}^{bv}|L_{n-j}^{\alpha+j-1}(t)|dt\Big)v^{\alpha+j-2}dv.$$

We split the v integral into two parts. Let

$$D_1 = \int_0^{(n-j)^{-1}} \Big(\int_a^{bv}|L_{n-j}^{\alpha+j-1}(t)|dt\Big)v^{\alpha+j-2}dv,$$

$$D_2 = \int_{(n-j)^{-1}}^1 \Big(\int_{bv}^{av}|L_{n-j}^{\alpha+j-1}(t)|dt\Big)v^{\alpha+j-2}dv.$$

Using the asymptotic properties of the Laguerre polynomials it is easy to see that

$$D_1 \leq C\int_0^{(n-j)^{-1}} \Big(\int_{av}^{bv} n^{\alpha+j-1}dt\Big)v^{\alpha+j-2}dv \leq Cn^{-1},$$

$$D_2 \leq C\int_{(n-j)^{-1}}^1 \Big(\int_{av}^{bv}\big(\frac{n}{t}\big)^{(\alpha+j-1)/2}(nt)^{-\frac{1}{4}}dt\Big)v^{\alpha+j-2}dv \leq Cn^{(\alpha+j)/2-3/4},$$

for large n. Combining the last two estimates we complete the proof of the Lemma. ∎

The above lemma enables us to extend the definition of T_α^β and τ_α^β to complex values of β. Using the multipliers m_λ^α and M_λ^α we want to effect a further reduction in the proof of the Proposition 7.1.1 and 7.1.2. Recall that m_λ^α is defined by

(7.2.8) $$m_\lambda^\alpha f = \sum_{k=0}^\infty \lambda(2k+1)(f,\mathcal{L}_k^\alpha)\mathcal{L}_k^\alpha$$

and M_λ^α is defined by the equation

(7.2.9) $$M_\lambda^\alpha f = \sum_{k=0}^\infty \lambda(2k+1)(f,\varphi_k^\alpha)\varphi_k^\alpha.$$

Let us take for λ the function defined by

(7.2.10) $$\lambda(2k+1) = \Big(\frac{\Gamma(k+\alpha+1+i\theta)}{\Gamma(k+\alpha+1)}\Big)^{\frac{1}{2}}$$

where we choose the branch of the square root which is equal to 1 for $\theta = 0$. We first verify that λ satisfies the following estimates.

Lemma 7.2.2. If we let λ as in (7.2.10) with $\alpha > -\frac{1}{2}$ then for each $j = 0, 1, 2, \ldots$ one has the estimates

$$\sup_{x>0} |\lambda^{(j)}(x) x^j| \leq C_j (1 + |\theta|)^j$$

for $\theta \in \mathbb{R}$, where C_j is independent of θ.

Proof: It is convenient to consider the function $m(x) = \lambda(2x)$. We will show that m satisfies the estimates of the Lemma. By Stirling's formula it follows that $|m(x)| \leq C$. Let $\psi(z)$ be the logarithmic derivative of $\Gamma(z)$, that is, $\psi(z) = \Gamma'(z)/\Gamma(z)$. If we let $u = x + \alpha + \frac{1}{2}$ so that $u > 0$ for $\alpha > -\frac{1}{2}$ we have

(7.2.11) $\qquad m'(x) = \frac{1}{2} m(x)(\psi(u + i\theta) - \psi(u)).$

Differentiating this identity j times we get

(7.2.12) $\quad m^{(j+1)}(x) = \frac{1}{2} \sum_{k=0}^{j} \binom{j}{k} m^{(j-k)}(x) \big(\psi^{(k)}(u + i\theta) - \psi^{(k)}(u) \big).$

In view of this identity it is enough to prove

(7.2.13) $\qquad \sup_{x>0} \big| \psi^{(k)}(u + i\theta) - \psi^{(k)}(u) \big| x^{k+1} \leq C |\theta|$

for every k.

To do this we need to use several formulas for the logarithmic derivative $\psi(z)$. For the formulas we use we refer to [EM1]. For $k = 0$ we use the formula

(7.2.14) $\qquad \psi(z) = \log z + \int_0^\infty \{(1 - e^t)^{-1} + t^{-1} - 1\} e^{-tz} dt$

valid for Re $z > 0$. This gives the estimate

$$|\psi(u + i\theta) - \psi(u)| \leq C |\theta| u^{-1}$$

which takes care of the case $k = 0$. When $k \geq 1$ we have the formula

(7.2.15) $\qquad \psi^{(k)}(u + i\theta) - \psi^{(k)}(u)$

$$= (-1)^{k+1} k! \sum_{j=0}^{\infty} \left(\frac{1}{u + i\theta + j} - \frac{1}{u + j} \right)$$

$$\times \left\{ \frac{1}{(u + i\theta + j)^k} + \frac{1}{(u + i\theta + j)^{k-1}(u+j)} + \cdots + \frac{1}{(u+j)^k} \right\}.$$

This gives us the estimate

$$|\psi^{(k)}(u+i\theta) - \psi^{(k)}(u)|x^{k+1} \tag{7.2.16}$$

$$\leq k!|\theta|x^{k+1}\sum_{j=0}^{\infty}(k+1)(u+j)^{-k-2} \leq C|\theta|$$

where C is a constant independent of x and θ. This completes the proof of the Lemma. ∎

Let us assume that α is a nonnegative integer. As the function λ verifies the estimates of the Lemma, it follows from Theorem 6.3.6 that m_λ^α is bouned on $L^p(\mathbb{R}_+)$, $1 < p < \infty$. Likewise, it follows from Theorem 6.4.4 that m_λ^α satisfies

$$\int_0^\infty |m_\lambda^\alpha f(x)|^p x^{p/4-1/2} dx \leq C \int_0^\infty |f(x)|^p x^{p/4-1/2} dx \tag{7.2.17}$$

which is equivalent to the boundedness of M_λ^α on $L^p(\mathbb{R}_+)$, $1 < p < \infty$. We now define the modified transplantation operators $\widetilde{T}_\alpha^\beta$ and $\widetilde{\tau}_\alpha^\beta$ by

$$\widetilde{T}_\alpha^\beta f(x) = \sum_{k=0}^{\infty} \left(\frac{\Gamma(k+\alpha+1)}{\Gamma(k+\alpha+1+i\theta)}\right)^{\frac{1}{2}} (f, \mathcal{L}_k^\beta)\mathcal{L}_k^\alpha, \tag{7.2.18}$$

$$\widetilde{\tau}_\alpha^\beta f(x) = \sum_{k=0}^{\infty} \left(\frac{\Gamma(k+\alpha+1)}{\Gamma(k+\alpha+1+i\theta)}\right)^{\frac{1}{2}} (f, \varphi_k^\beta)\varphi_k^\alpha. \tag{7.2.19}$$

It then follows from the definition of λ that

$$m_\lambda^\alpha \widetilde{T}_\alpha^\beta f = T_\alpha^\beta f, \quad M_\lambda^\alpha \widetilde{\tau}_\alpha^\beta f = \tau_\alpha^\beta f. \tag{7.2.20}$$

As the operators m_λ^α and M_λ^α are bounded on $L^p(\mathbb{R}_+)$ for α a nonnegative integer, Proposition 7.1.1 will be proved once we establish the following.

Proposition 7.2.1. Let $\alpha = 0, 1, 2, \ldots$, $f \in L^p(\mathbb{R}_+)$ and $1 < p < \infty$. Then one has for $k = 0, 2$

(i) $\|\widetilde{T}_\alpha^{\alpha+k+i\theta} f\|_p \leq M(\theta)\|f\|_p$,

(ii) $\|\widetilde{\tau}_\alpha^{\alpha+k+i\theta} f\|_p \leq M(\theta)\|f\|_p$.

We will actually establish more. Let us denote by $\|f\|_{p,w}$ the weighted norm

(7.2.21) $$\|f\|_{p,w} = \left(\int_0^\infty |f(x)|^p x^{p/4-1/2} dx \right)^{\frac{1}{p}}.$$

Then we know that (ii) of the above proposition is equivalent to the weighted estimate

(7.2.22) $$\|\widetilde{T}_\alpha^{\alpha+k+i\theta} f\|_{p,w} \leq M(\theta) \|f\|_{p,w}.$$

Thus Propositions 7.1.2 and 7.2.1 will be established once we have the following proposition.

Proposition 7.2.2.

(i) If $\alpha \geq 0$ then one has $\|\widetilde{T}_\alpha^{\alpha+k+i\theta} f\|_p \leq M(\theta) \|f\|_p$, $1 < p < \infty$ for $k = 0, 2$. If $-1 < \alpha < 0$ the estimate is valid with $k = 2$ and p satisfying $\left(1 + \frac{\alpha}{2}\right)^{-1} < p < -\frac{2}{\alpha}$.

(ii) If $\alpha \geq 0$ then one also has the weighted inequality $\|\widetilde{T}_\alpha^{\alpha+k+i\theta} f\|_{p,w} \leq M(\theta) \|f\|_{p,w}$, $1 < p < \infty$ for $k = 0, 2$. If $\alpha \geq -\frac{1}{2}$ the same is true with $k = 2$ and $1 < p < \infty$.

Observe that when $\theta = 0$, $\widetilde{T}_\alpha^{\alpha+2} = T_\alpha^{\alpha+2}$ so that Proposition 7.1.2 follows from the above proposition. Thus the crux of the matter lies in proving the above proposition. The proposition will be established by obtaining integral representations for the operators $\widetilde{T}_\alpha^{\alpha+k+i\theta}$. We also need a general multiplier theorem in order to deal with $\widetilde{T}_\alpha^{\alpha+2+i\theta}$. In the next section we state and prove some preliminary Lemmas which are needed in obtaining the integral representations.

7.3 Some Preliminary Lemmas

In order to establish certain integral representations of $\widetilde{T}_\alpha^{\alpha+k+i\theta}$ we have to use some properties of kernels defined by Bessel functions. In the study of $\widetilde{T}_\alpha^{\alpha+i\theta}$ we will come across the kernel $X_\gamma(w,t)$ which is defined by

(7.3.1) $$X_\gamma(w,t) = \int_0^\infty J_\alpha(ws) J_\alpha(ts) e^{-\gamma s^2/2} ds$$

where $0 < \gamma < 1$. For this kernel we need the following fact.

Lemma 7.3.1. Let $\alpha > -1$, $0 < \gamma < 1$ and $t > 0$. Then for $g \in C_0^\infty(\mathbb{R}_+)$ one has

$$\lim_{\gamma \to 0} \int_0^\infty g(w) X_\gamma(w,t) dw = t^{-1} g(t).$$

Proof: The proof needs certain formulas which are proved in [EM1]. First of all one has the formula

$$(7.3.2) \qquad X_\gamma(w,t) = \frac{1}{\gamma} e^{-\frac{(w^2+t^2)}{2\gamma}} I_\alpha\left(\frac{wt}{\gamma}\right)$$

where I_α is the modified Bessel function. If we let

$$(7.3.3) \qquad W_\gamma(t) = \int_0^\infty X_\gamma(w,t) dw$$

then we also have the formula

$$(7.3.4) \qquad W_\gamma(t) = \left(\frac{\pi}{2\gamma}\right)^{\frac{1}{2}} e^{-\frac{t^2}{4\gamma}} I_{\frac{\alpha}{2}}\left(\frac{t^2}{4\gamma}\right).$$

By the asymptotic formula

$$(7.3.5) \qquad I_\alpha(z) = (2\pi z)^{-\frac{1}{2}} \left\{ e^z (1 + 0(|z|^{-1})) + i e^{-z + \alpha i \pi} (1 + 0(|z|^{-1})) \right\}$$

valid for $-\frac{\pi}{2} < \arg z < \frac{3\pi}{2}$ we see that $\lim_{\gamma \to 0} W_\gamma(t) = t^{-1}$. If we let $Z_\gamma(w,t) = W_\gamma(t)^{-1} X_\gamma(w,t)$ it also follows that

$$(7.3.6) \qquad Z_\gamma(w,t) \leq C \gamma^{-\frac{1}{2}} e^{-\frac{(w-t)^2}{2\gamma}}$$

for $a \leq t$, $w \leq b$, where C is a constant independent of t, w and γ. This proves that $Z_\gamma(w,t)$ is a summability kernel and hence the Lemma. ∎

In the estimation of $\widetilde{T}_\alpha^{\alpha+2+i\theta} f$ we will come across another kernel $R_\lambda(u,x)$ which is defined by

$$(7.3.7) \qquad R_\lambda(u,x) = \int_0^\infty J_{\alpha+3}(us) J_\alpha(xs) s^{-\lambda} ds$$

where u, x, λ are all positive. This can be expressed in terms of the hypergeometric function $_2F_1$. In fact, one has the formulas

$$(7.3.8) \qquad R_\lambda(u,x) = \frac{x^\alpha \Gamma(\alpha + 2 - \frac{\lambda}{2})}{2^\lambda u^{\alpha+1-\lambda} \Gamma(\alpha+1) \Gamma(2 + \frac{\lambda}{2})}$$
$$\times \; _2F_1\left(\alpha + 2 - \frac{\lambda}{2}, -1 - \frac{\lambda}{2}; \alpha + 1; \left(\frac{x}{u}\right)^2\right)$$

for $x < uA$, and

(7.3.9) $$R_\lambda(u,x) = \frac{u^{\alpha+3}\Gamma(\alpha+2-\frac{\lambda}{2})}{2^\lambda x^{\alpha+4-\lambda}\Gamma(\alpha+4)\Gamma(\frac{\lambda}{2}-1)}$$
$$\times {}_2F_1\left(\alpha+2-\tfrac{\lambda}{2}, 2-\tfrac{\lambda}{2}; \alpha+4; \left(\tfrac{u}{x}\right)^2\right)$$

for $x > u$.

These formulas are valid for $(2\alpha + 4) > \lambda > -1$. Using the above expressions it is not difficult to establish the following lemma.

Lemma 7.3.2. *Let $x > 0$, $b < \infty$ and $0 < \lambda < 2(\alpha + 1)$. Then one has the estimates*

(i) $|R_\lambda(u,x)| \leq C$, $0 \leq u \leq b$, $x > u$

(ii) $|R_\lambda(u,x)| \leq C \log\{1 + (1 - (\frac{x}{u})^2)^{-1}\}$, $0 \leq u \leq b$, $u > x$.

We have the limits

(iii) $\lim_{\lambda \to 0} R_\lambda(u,x) = 0$, $x > u$.

(iv) $\lim_{\lambda \to 0} R_\lambda(u,x) = \frac{(\alpha+1)x^\alpha}{u^{\alpha+1}}\left\{1 - \frac{\alpha+2}{\alpha+1}(\frac{x}{u})^2\right\}$, $u > x$.

Proof: Since ${}_2F_1(\alpha,\beta;\gamma;z)$ is a continuous function of (α,β) for fixed z and γ, the limits (iii) and (iv) follow from (7.3.8) and (7.3.9) and the fact that

$${}_2F_1\left(\alpha+2, -1; \alpha+1; \left(\tfrac{x}{u}\right)^2\right) = \left(1 - \frac{\alpha+2}{\alpha+1}\left(\tfrac{x}{u}\right)^2\right).$$

To prove (i) when $x > u$ we use the integral formula

$${}_2F_1\left(\alpha+2-\frac{\lambda}{2}, 2-\frac{\lambda}{2}; \alpha+4; \left(\frac{u}{x}\right)^2\right)$$

$$= \frac{\Gamma(\alpha+4)}{\Gamma(2-\frac{\lambda}{2})\Gamma(\alpha+2+\frac{\lambda}{2})}$$

$$\times \int_0^1 t^{1-\frac{\lambda}{2}}(1-t)^{\alpha+1+\frac{\lambda}{2}}\left(1-t\left(\frac{u}{x}\right)^2\right)^{-\alpha-2+\frac{\lambda}{2}}dt$$

$$\leq \frac{\Gamma(\alpha+4)}{\Gamma(2-\frac{\lambda}{2})\Gamma(\alpha+2+\frac{\lambda}{2})} B\left(2-\frac{\lambda}{2}, \lambda\right)$$

$$= \frac{\Gamma(\alpha+4)\Gamma(\lambda)}{\Gamma(2+\frac{\lambda}{2})\Gamma(\alpha+2+\frac{\lambda}{2})}.$$

This and (7.39) prove (i). When $x < u$ we use the formula

$$_2F_1\left(\alpha + 2 - \frac{\lambda}{2}, -1 - \frac{\lambda}{2}; \alpha + 1; \left(\frac{x}{u}\right)^2\right)$$

$$= \frac{ie^{i\pi(1-\frac{\lambda}{2})}\Gamma(\alpha+1)\Gamma(2-\frac{\lambda}{2})}{2\pi\Gamma(\alpha+2-\frac{\lambda}{2})}$$

$$\times \int_0^{(1+)} t^{\alpha+1-\frac{\lambda}{2}}(1-t)^{-2+\frac{\lambda}{2}}\left(1 - t\left(\frac{x}{u}\right)^2\right)^{1+\frac{\lambda}{2}} dt$$

when $\alpha + 2 - \frac{\lambda}{2} > 0$ and $-1 + \frac{\lambda}{2} \neq 1, 2, \ldots$. Here the integral is taken along a contour which starts from the origin, encircles the point 1 once counterclockwise and returns to the origin. All singularities of the integrand except 1 are outside the contour. This formula leads to the estimate (ii). ∎

A general multiplier theorem due to Butzer, Nessel and Trebels [BNT] is also needed in the study of $\tilde{T}_\alpha^{\alpha+2+i\theta}$. To state this result one has to recall the definition of a bounded quasiconvex sequence. A sequence $\lambda = (\lambda(k))$ is said to be a bounded quasi-convex sequence if

$$\|\lambda\|_{bqc} = \sum_{k=0}^{\infty}(k+1)|\Delta^2\lambda(k)| + \lim_{k\to\infty}|\lambda(k)|$$

is finite where $\Delta^2\lambda(k) = \lambda(k) - 2\lambda(k+1) + \lambda(k+2)$. For such a sequence λ one defines the multiplier $\tilde{m}_\lambda^\alpha$ by

(7.3.10) $$\tilde{m}_\lambda^\alpha f(x) = \sum_{k=0}^{\infty} \lambda(k)(f, \mathcal{L}_k^\alpha)\mathcal{L}_k^\alpha.$$

Note that this is just a variant of m_λ^α. Regarding the boundedness of $\tilde{m}_\lambda^\alpha$ we have the following result:

Theorem 7.3.1. (*Butzer-Nessel-Trebels*) *If λ is a bounded quasi-convex sequence then for $1 < p < \infty$*

(i) $\|\tilde{m}_\lambda^\alpha f\|_p \leq C\|\lambda\|_{bqc}\|f\|_p$,

(ii) $\|\tilde{m}_\lambda^\alpha f\|_{p,w} \leq C\|\lambda\|_{bqc}\|f\|_{p,w}$.

Proof: The proof is based on the following identity. Let $\mu(k) = \lambda(k) - \lambda(\infty)$ where $\lambda(\infty) = \lim_{k\to\infty}\lambda(k)$. If $C_N^1 f$ is the Cesàro means studied in Section 6.3 then we have the identity

(7.3.11) $$\tilde{m}_\lambda^\alpha f = \sum_{k=0}^{\infty}(k+1)\Delta^2\mu(k)C_k^1 f + \lambda(\infty)f.$$

This can be verified using the fact that

$$(7.3.12) \qquad C_N^1 f = \sum_{k=0}^{N} \left(1 - \frac{k}{N+1}\right)(f, \mathcal{L}_k^\alpha)\mathcal{L}_k^\alpha.$$

It follows from (7.3.11) and the boundedness of C_N^1 (Theorem 6.3.2) that

$$(7.3.13) \qquad \|\widetilde{m}_\lambda^\alpha f\|_p \leq C\|\lambda\|_{bqc}\|f\|_p.$$

If we take $\beta = 1/4 - 1/2p$, it satisfies the conditions given in Theorem 6.3.2 and hence the weighted inequality (ii) is also valid. ∎

In Section 7.4 we need to apply the above theorem with the sequence λ defined by

$$(7.3.14) \quad \lambda(k) = \alpha + \tfrac{3}{2} + i\theta - \tfrac{1}{4}\Big\{(k+\alpha+2+i\theta)^{\frac{1}{2}}(k+\alpha+1+i\theta)^{\frac{1}{2}}$$
$$+ \left(k+\alpha+\tfrac{3}{2}+i\theta\right)\Big\}^{-1}.$$

To use the theorem we need to know if λ is a bounded quasi-convex sequence. That is the content of the next lemma.

Lemma 7.3.3. *If $\alpha > -1$ then $\|\lambda\|_{bqc} \leq C(1 + |\theta|)$ where C depends only on α.*

Proof: Let $\eta(x) = (x+a)^{\frac{1}{2}}(x+a+1)^{\frac{1}{2}}$, $a = \alpha + 1 + i\theta$, where the branch is so chosen that $\eta(x)$ is positive for $\theta = 0$. Let $\lambda(x) = (\eta(x) + x + b)^{-1}$ with $b = (a + \tfrac{1}{2})$. We then have $\lambda''(x) = \eta(x)^{-3}$. Thus $|\Delta^2 \lambda(k)| \leq 2\max\{|\lambda''(x)| : k \leq x \leq k+2\} \leq 2(k+\alpha+1)^{-3}$. Since $\lim_{k\to\infty} \lambda(k) = \alpha + \tfrac{3}{2} + i\theta$, the Lemma is proved. ∎

With these preparatory lemmas we can now proceed to the estimation of $\widetilde{T}_\alpha^{\alpha+k+i\theta} f$. The two results we use in the estimation are the boundedness of Calderon–Zygmund singular integral operators on L^p spaces and the L^p boundedness of the Hardy operator

$$(7.3.15) \qquad Hf(x) = \int_x^\infty f(t) t^{-1} dt$$

namely, $\|Hf\|_p \leq C\|f\|_p$, $1 < p < \infty$. We also need weighted versions of these two results. If T is a Calderon–Zygmund singular integral operator and $w \in A_p$, the Muckenhoupt class and $1 < p < \infty$ then

$$(7.3.16) \qquad \|Tf\|_{p,w} \leq C\|f\|_{p,w}.$$

When $1 < p < \infty$ and $\beta + 1 > 0$ we also have

(7.3.17) $$\Big(\int_0^\infty |Hf(x)|^p x^\beta \, dx \Big)^{\frac{1}{p}} \leq C \Big(\int_0^\infty |f(x)|^p x^\beta \, dx \Big)^{\frac{1}{p}}.$$

The functions $\widetilde{T}_\alpha^{\alpha+k+i\theta} f$ will be estimated in the following two sections.

7.4 Estimating $\widetilde{T}_\alpha^{\alpha+i\theta} f$

Let $\varepsilon > 0$ and define $G_\varepsilon^\alpha f$ by the formula

(7.4.1) $$G_\varepsilon^\alpha f(x) = \sum_{k=0}^\infty \Big(\frac{\Gamma(k+\alpha+1)}{\Gamma(k+\alpha+1+\varepsilon+i\theta)} \Big)^{\frac{1}{2}} (f, \mathcal{L}_k^{\alpha+\varepsilon+i\theta}) \mathcal{L}_k^\alpha$$

where we are taking the branch of the square root which is positive for $\theta = 0$. In view of the estimates (7.1.12) it follows that for each $x > 0$

(7.4.2) $$\lim_{\varepsilon \to 0} G_\varepsilon^\alpha f(x) = \widetilde{T}_\alpha^{\alpha+i\theta} f(x).$$

We shall show that

(7.4.3) $$\|G_\varepsilon^\alpha f\|_p \leq M(\theta)(\|f(x) x^{\varepsilon/2}\|_p + \|f(x) x^{-\varepsilon/2}\|_p)$$

for $\alpha \geq 0$, $1 < p < \infty$, $0 < \varepsilon < 1$, $-\infty < \theta < \infty$ and $f \in C_0^\infty(\mathbb{R}_+)$. We shall also prove the weighted inequality

(7.4.4) $$\|G_\varepsilon^\alpha f\|_{p,w} \leq M(\theta)(\|f(x) x^{\varepsilon/2}\|_{p,w} + \|f(x) x^{-\varepsilon/2}\|_{p,w})$$

for G_ε^α. By letting ε tend to 0, using Fatou's lemma and dominated convergence theorem we get

(7.4.5) $$\|\widetilde{T}_\alpha^{\alpha+i\theta} f\|_p \leq M(\theta) \|f\|_p, \quad \alpha \geq 0, \quad 1 < p < \infty$$

and

(7.4.6) $$\|\widetilde{T}_\alpha^{\alpha+i\theta} f\|_{p,w} \leq M(\theta) \|f\|_{p,w}, \quad \alpha \geq 0, \quad 1 < p < \infty.$$

In order to prove (7.4.3) and (7.4.4) we first establish an integral representation for $G_\varepsilon^\alpha f$. It is for this purpose we need Lemma 7.3.1. We prove the following proposition.

Proposition 7.4.1. Let $\alpha > -1$, $\varepsilon > 0$ and $\theta \in \mathbb{R}$. Then

$$G_\varepsilon^\alpha f(x) = \frac{e^{x/2}}{\Gamma(\varepsilon + i\theta)} \int_0^1 v^{\alpha/2-1}(1-v)^{\varepsilon-1+i\theta} f\Big(\frac{x}{v}\Big) e^{-\frac{x}{2v}} \Big(\frac{x}{v}\Big)^{\frac{\varepsilon+i\theta}{2}} dv.$$

Proof: Let $0 < r < 1$ and define $G_{\varepsilon,r}^\alpha f$ by

(7.4.7) $\qquad G_{\varepsilon,r}^\alpha f = \sum_{k=0}^\infty \Big(\frac{\Gamma(k+\alpha+1)}{\Gamma(k+\alpha+1+\varepsilon+i\theta)}\Big)^{\frac{1}{2}} r^k (f, \mathcal{L}_k^{\alpha+\varepsilon+i\theta}) \mathcal{L}_k^\alpha$

which we write in the form

(7.4.8) $\qquad G_{\varepsilon,r}^\alpha f(x) = \int_0^\infty G_{\varepsilon,r}^\alpha(x,y) f(y) e^{-\frac{1}{2}y} y^{\frac{\alpha+\varepsilon+i\theta}{2}} dy$

with the obvious definition of $G_{\varepsilon,r}^\alpha(x,y)$. Now we use the formula (see [GS])

(7.4.9) $\qquad L_k^\alpha(x) e^{-x} x^{\alpha/2} = \frac{1}{k!} \int_0^\infty e^{-t} t^{k+\alpha/2} J_\alpha(2(tx)^{\frac{1}{2}}) dt$

to rewrite $G_{\varepsilon,r}^\alpha(x,y)$ in the form

(7.4.10) $\qquad G_{\varepsilon,r}^\alpha(x,y)$

$$= e^{x/2} \int_0^\infty \Big(\sum_{k=0}^\infty \frac{(rt)^k}{\Gamma(k+\alpha+1+\varepsilon+i\theta)} L_k^{\alpha+\varepsilon+i\theta}(y) \Big)$$
$$\times J_\alpha(2(tx)^{\frac{1}{2}}) e^{-t} t^{\alpha/2} dt$$

In this expression we want to write $L_k^{\alpha+\varepsilon+i\theta}$ in terms of L_k^α. Using the formula (7.2.4) we get

(7.4.11) $\qquad L_k^{\alpha+\varepsilon+i\theta}(y)$

$$= \frac{\Gamma(k+\alpha+\varepsilon+1+i\theta)}{\Gamma(\varepsilon+i\theta)\Gamma(k+\alpha+1)} \int_0^1 v^\alpha (1-v)^{\varepsilon-1+i\theta} L_k^\alpha(vy) dv.$$

Using this formula in (7.4.10) we have

(7.4.12) $\qquad G_{\varepsilon,r}^\alpha(x,y) = \frac{e^{x/2}}{\Gamma(\varepsilon+i\theta)} \int_0^\infty \int_0^1 \Big(\sum_{k=0}^\infty \frac{(rt)^k}{\Gamma(k+\alpha+1)} L_k^\alpha(vy) \Big)$

$$\times v^\alpha (1-v)^{\varepsilon-1+i\theta} J_\alpha(2(tx)^{\frac{1}{2}}) e^{-t} t^{\alpha/2} dv dt.$$

Finally, using the generating function identity

$$(7.4.13) \qquad \sum_{k=0}^{\infty} \frac{w^k}{\Gamma(k+\alpha+1)} L_k^\alpha(y) = e^w (yw)^{-\alpha/2} J_\alpha(2(yw)^{\frac{1}{2}})$$

we obtain the expression

$$(7.4.14) \qquad G_{\varepsilon,r}^\alpha(x,y) = \frac{e^{x/2}(ry)^{-\alpha/2}}{\Gamma(\varepsilon+i\theta)} \int_0^\infty \int_0^1 (1-v)^{\varepsilon-1+i\theta} J_\alpha(2(tx)^{\frac{1}{2}})$$
$$\times J_\alpha(2(rtvy)^{\frac{1}{2}}) e^{rt} e^{-t} dv dt.$$

Now a change of variables in the above expression gives us

$$(7.4.15) \qquad G_{\varepsilon,r}^\alpha(x^2,y^2) = \frac{2e^{x^2/2} y^{-\alpha} r^{-\alpha-1}}{\Gamma(\varepsilon+i\theta)} \int_0^\infty \int_0^1 u^{\alpha+1}\left(1-\frac{u^2}{r}\right)^{\varepsilon-1+i\theta}$$
$$\times J_\alpha(\sqrt{2}xs) J_\alpha(\sqrt{2}uys) e^{-(1-r)\frac{s^2}{2}} s\, du\, ds.$$

Using the asymptotic properties of the Bessel function it is not difficult to show that the order of the integration can be changed. That leads to the formula

$$(7.4.16) \qquad G_{\varepsilon,r}^\alpha(x^2,y^2) = \frac{2e^{x^2/2} r^{-\alpha-1} y^{-\alpha}}{\Gamma(\varepsilon+i\theta)} \int_0^1 u^{\alpha+1}\left(1-\frac{u^2}{r}\right)^{\varepsilon-1+i\theta}$$
$$X_{1-r}(\sqrt{2}x,\sqrt{2}uy) du$$

where $X_\gamma(w,t)$ is defined in (7.3.1). Therefore, we have proved

$$(7.4.17) \qquad G_{\varepsilon,r}^\alpha f(x^2) = 2\int_0^\infty G_{\varepsilon,r}^\alpha(x^2,y^2) f(y^2) e^{-\frac{1}{2}y^2} y^{\alpha+\varepsilon+1+i\theta} dy$$
$$= 4\frac{e^{x^2/2} r^{-\alpha-1}}{\Gamma(\varepsilon+i\theta)} \int_0^\infty \int_0^1 u^{\alpha+1}\left(1-\frac{u^2}{\gamma}\right)^{\varepsilon-1+i\theta}$$
$$X_{1-r}(\sqrt{2}x,\sqrt{2}uy) f(y^2) e^{-\frac{1}{2}y^2} y^{\varepsilon+1+i\theta} du\, dy.$$

Taking limit as r tends to 1 and using Lemma 7.3.1 we obtain

$$(7.4.18) \qquad G_\varepsilon^\alpha f(x^2) = \frac{2e^{x^2/2}}{\Gamma(\varepsilon+i\theta)} \int_0^1 u^{\alpha-1}(1-u^2)^{\varepsilon-1+i\theta}$$
$$\times f\left(\left(\frac{x}{u}\right)^2\right) e^{-\frac{1}{2}(\frac{x}{u})^2}\left(\frac{x}{u}\right)^{\varepsilon+i\theta} du.$$

A change of variable now completes the proof of the proposition. ∎

Having obtained an integral representation for $G_\varepsilon^\alpha f$ we can now establish the estimates (7.4.3) and (7.4.4). Assume $\alpha \geq 0$ and $0 < \varepsilon < 1$. Making a change of variables we can write $G_\varepsilon^\alpha f(x) = I_\varepsilon^\alpha f(x) + J_\varepsilon^\alpha f(x)$ where

$$(7.4.19) \qquad I_\varepsilon^\alpha f(x) = \frac{1}{\Gamma(\varepsilon + i\theta)} \int_x^\infty \frac{f(t)}{t} e^{-(t-x)/2} t^{(\varepsilon+i\theta)/2}$$

$$\times \left(1 - \frac{x}{t}\right)^{\varepsilon-1+i\theta} dt,$$

$$(7.4.20) \qquad J_\varepsilon^\alpha f(x) = \frac{1}{\Gamma(\varepsilon + i\theta)} \int_x^\infty \frac{f(t)}{t} e^{-(t-x)/2} t^{(\varepsilon+i\theta)/2}$$

$$\times \left(\left(\frac{x}{t}\right)^{\alpha/2} - 1\right)\left(1 - \frac{x}{t}\right)^{\varepsilon-1+i\theta} dt.$$

Since $\alpha \geq 0$ it follows that $\left(\left(\frac{x}{t}\right)^{\alpha/2} - 1\right)\left(1 - \frac{x}{t}\right)^{\varepsilon-1+i\theta}$ is bounded by a constant C for $t > x$. It follows that

$$(7.4.21) \qquad |J_\varepsilon^\alpha f(x)| \leq \frac{C}{|\Gamma(\varepsilon + i\theta)|} \int_x^\infty |f(t)| t^{\varepsilon/2 - 1} dt.$$

We remark that we have the estimate $|\Gamma(\varepsilon + i\theta)|^{-1} \leq A(1 + |\theta|)e^{\frac{\pi}{2}|\theta|}$ where A is an absolute constant. Therefore, Hardy's inequality proves that

$$(7.4.22) \qquad \|J_\varepsilon^\alpha f\|_p \leq M(\theta)\|f(x)x^{\varepsilon/2}\|_p,$$

$$(7.4.23) \qquad \|J_\varepsilon^\alpha f\|_{p,w} \leq M(\theta)\|f(x)x^{\varepsilon/2}\|_{p,w}.$$

This takes care of $J_\varepsilon^\alpha f$. We next consider $I_\varepsilon^\alpha f(x)$. We extend $f \in C_0^\infty(\mathbb{R}_+)$ into the function on $\mathbb{R}$ which coincides with f on $\mathbb{R}_+$ and vanishes on $(-\infty, 0)$. We denote this function by $\widetilde{f}$. We define

$$(7.4.24) \qquad \widetilde{I}_\varepsilon^\alpha f(x) = \int_{-\infty}^\infty f(t) Q(x-t) dt$$

where Q is the kernel defined by

$$(7.4.25) \qquad Q(u) = \frac{1}{\Gamma(\varepsilon + i\theta)} e^{-|u|/2} |u|^{\varepsilon - 1 + i\theta} \chi_{(-\infty, 0)}(u).$$

We note that $I_\varepsilon^\alpha f(x) = \widetilde{I}_\varepsilon^\alpha \widetilde{f}_\varepsilon(x)$ where $\widetilde{f}_\varepsilon$ is the function

$$\widetilde{f}_\varepsilon(t) = \widetilde{f}(t) t^{-(\varepsilon + i\theta)/2}.$$

We shall show that $\widetilde{I}_\varepsilon^\alpha$ is a singular integral operator. To see this we need to calculate the Fourier transform of the function Q. Using some well known formulas in [EM2] we have

(7.4.26) $\quad \hat{Q}(y) = (2\pi)^{-\frac{1}{2}} \left(\frac{1}{4} + y^2\right)^{-(\varepsilon+i\theta)/2} e^{i\varepsilon \tan^{-1} 2y} e^{-\theta \tan^{-1} 2y}.$

This gives the estimate

(7.4.27) $\quad |\hat{Q}(y)| \leq 2 e^{\pi|\theta|/2} = B_1.$

We also see that

(7.4.28) $\quad \left|\dfrac{d}{du} Q(y)\right| \leq A \dfrac{(1+|\theta|)}{|\Gamma(\varepsilon + i\theta)|} u^{-2}$

$\qquad \leq A(1+|\theta|) e^{\frac{\pi}{2}|\theta|} u^{-2} = B_2 u^{-2}.$

Thus $\widetilde{I}_\varepsilon^\alpha$ is a Calderon–Zygmund singular integral operator and hence is bounded on $L^p(\mathbb{R})$, $1 < p < \infty$ and also satisfies a weighted inequality when the weight $w \in A_p$. Keeping track of the constants one gets

(7.4.29) $\quad \|\widetilde{I}_\varepsilon^\alpha f\|_p \leq M(\theta) \|f\|_p,$

(7.4.30) $\quad \|\widetilde{I}_\varepsilon^\alpha f\|_{p,w} \leq M(\theta) \|f\|_{p,w}.$

The constant $M(\theta)$ depends only on the constants B_1 and B_2 and hence is of admissible growth. This leads to the estimates

(7.4.31) $\quad \|I_\varepsilon^\alpha f\|_p \leq M(\theta) \|f(x) x^{-\varepsilon/2}\|_p,$

(7.4.32) $\quad \|I_\varepsilon^\alpha f\|_{p,w} \leq M(\theta) \|f(x) x^{-\varepsilon/2}\|_{p,w}.$

Thus we have established the inequalities (7.4.3) and (7.4.4) which will prove (7.4.5) and (7.4.6).

Thus we have taken care of $\widetilde{T}_\alpha^{\alpha+i\theta} f$ when $\alpha \geq 0$. In the next section we consider $\widetilde{T}_\alpha^{\alpha+2+i\theta} f$ when $\alpha > -1$.

7.5 Estimating $\widetilde{T}_\alpha^{\alpha+2+i\theta} f$

We first write $\widetilde{T}_\alpha^{\alpha+2+i\theta} f$ as a sum of two terms one of which can be estimated using the expression for $G_\varepsilon^\alpha f$ and the multiplier theorem of Butzer, Nessel and Trebels. From the definition we have

(7.5.1) $$\widetilde{T}_\alpha^{\alpha+2+i\theta} f = \sum_{k=0}^\infty \left(\frac{\Gamma(k+\alpha+1)}{\Gamma(k+\alpha+1+i\theta)} \right)^{\frac{1}{2}} (f, \mathcal{L}_k^{\alpha+2+i\theta}) \mathcal{L}_k^\alpha.$$

If we let

(7.5.2) $$\sigma(k) = \left(\frac{\Gamma(k+\alpha+3+i\theta)}{\Gamma(k+\alpha+1)} \right)^{\frac{1}{2}}$$

it is not difficult to show that

(7.5.3) $$\left(\frac{\Gamma(k+\alpha+1)}{\Gamma(k+\alpha+1+i\theta)} \right)^{\frac{1}{2}} = (\lambda(k) + k)\sigma(k)^{-1}$$

where $\lambda(k)$ is defined by (7.3.14). So, we can write

$$\widetilde{T}_\alpha^{\alpha+2+i\theta} f = U^\alpha f + V^\alpha f$$

with

(7.5.4) $$U^\alpha f = \sum_{k=0}^\infty \lambda(k)\sigma(k)^{-1}(f, \mathcal{L}_k^{\alpha+2+i\theta})\mathcal{L}_k^\alpha,$$

(7.5.5) $$V^\alpha f = \sum_{k=0}^\infty k\sigma(k)^{-1}(f, \mathcal{L}_k^{\alpha+2+i\theta})\mathcal{L}_k^\alpha.$$

We first deal with $U^\alpha f$. It is clear that one has the relation

(7.5.6) $$U^\alpha f = \widetilde{m}_\lambda^\alpha G_2^\alpha f.$$

By Lemma 7.3.3 the sequence $\lambda = (\lambda(k))$ is a bounded quasi-convex one and hence in view of Theorem 7.3.1 we have

(7.5.7) $$\|U^\alpha f\|_p \leq M(\theta)\|G_2^\alpha f\|_p,$$

(7.5.8) $$\|U^\alpha f\|_{p,w} \leq M(\theta)\|G_2^\alpha f\|_{p,w}.$$

So, it remains to estimate $G_2^\alpha f$. Using the integral representation of Proposition 7.4.1 and applying Minkowski's integral inequality we see that

$$(7.5.9) \qquad \|G_2^\alpha f\|_p \leq \frac{1}{|\Gamma(2+i\theta)|} \int_0^1 v^{\alpha/2-2}(1-v)$$
$$\times \left\{\int_0^\infty |f(\frac{x}{v})e^{x(1-\frac{1}{v})}x|^p dx\right\}^{\frac{1}{p}} dv$$
$$\leq \frac{1}{|\Gamma(2+i\theta)|} \int_0^1 v^{\alpha/2+1/p-1}(1-v)$$
$$\times \left\{\int_0^\infty |f(t)e^{t(v-1)/2}|^p dt\right\}^{\frac{1}{p}} dv.$$

Since $e^{t(v-1)/2}t \leq 2e^{-1}(1-v)^{-1}$ for $t > 0$ and $0 < v < 1$ it follows that

$$(7.5.10) \qquad \|G_2^\alpha f\|_p \leq 2e^{-1}|\Gamma(2+i\theta)|^{-1}\|f\|_p \int_0^1 v^{\alpha/2+1/p-1} dv.$$

The last integral is finite if $\alpha \geq 0$, $1 \leq p < \infty$ or $-1 < \alpha < 0$, $p < -\frac{2}{\alpha}$.

This proves that $\|U^\alpha f\|_p \leq M(\theta)\|f\|_p$ for $1 \leq p < \infty$ if $\alpha \geq 0$ and for $p < -\frac{2}{\alpha}$ if $-1 < \alpha < 0$. The estimation of $\|G_2^\alpha\|_{p,w}$ is similar. As before, Minkowski's integral inequality gives

$$(7.5.11) \qquad \|G_2^\alpha f\|_{p,w} \leq M(\theta)\|f\|_{p,w} \int_0^1 v^{\alpha/2+1/2p+1/4-1} dv.$$

The last integral is finite for $\alpha \geq -\frac{1}{2}$. Hence we also have $\|U^\alpha f\|_{p,w} \leq M(\theta)\|f\|_{p,w}$ for $1 < p < \infty$, $\alpha \geq -\frac{1}{2}$. This takes care of $U^\alpha f$.

We next consider the term $V^\alpha f$. To estimate this we take $\varepsilon > 0$ and define

$$(7.5.12) \qquad V_\varepsilon^\alpha f = \sum_{k=0}^\infty k\left(\frac{\Gamma(k+\alpha+1)}{\Gamma(k+\alpha+3+\varepsilon+i\theta)}\right)^{\frac{1}{2}}(f, \mathcal{L}_k^{\alpha+2+\varepsilon+i\theta})\mathcal{L}_k^\alpha.$$

As $\varepsilon \to 0$, $V_\varepsilon^\alpha f(x) \to V^\alpha f(x)$ for $x > 0$; therefore, it is enough to get estimates for the L^p norms of $V_\varepsilon^\alpha f$. We shall prove that

$$(7.5.13) \qquad \|V_\varepsilon^\alpha f\|_p \leq M(\theta)(\|f(x)x^{\varepsilon/2}\|_p + \|f(x)x^{-\varepsilon/2}\|_p)$$

for $\alpha \geq 0$, $1 < p < \infty$ or $-1 < \alpha < 0$, $p > -\frac{2}{\alpha}$. We shall also prove the weighted inequality

(7.5.14) $\quad \|V_\varepsilon^\alpha f\|_{p,w} \leq M(\theta)(\|f(x)x^{\varepsilon/2}\|_{p,w} + \|f(x)x^{-\varepsilon/2}\|_{p,w})$

for $\alpha \geq -\frac{1}{2}$, $1 < p < \infty$. Taking $\varepsilon \to 0$ we get the desired inequality for $V^\alpha f$.

We observe that $V_\varepsilon^\alpha f$ is defined like $G_\varepsilon^\alpha f$ with ε replaced by $\varepsilon + 2$ and an extra factor under the summation sign. Proceeding as in the case of $G_\varepsilon^\alpha f$ we can show that

(7.5.15) $\quad V_\varepsilon^\alpha f(x) = \dfrac{e^{x/2}}{\Gamma(\varepsilon + i\theta)} \int_0^\infty \int_0^\infty \int_0^1 v^{\alpha+2}(1-v)^{\varepsilon-1+i\theta}$

$$\times \Big(\sum_{k=0}^\infty \dfrac{kt^k}{\Gamma(k+\alpha+2+1)} L_k^{\alpha+2}(vy) \Big)$$

$$\times J_\alpha(2(tx)^{\frac{1}{2}})e^{-t}t^{\alpha/2}f(y)e^{-y/2}y^{\frac{\alpha+2+\varepsilon+i\theta}{2}} dv dy dt.$$

The formula (7.4.13) leads to the formula

(7.5.16) $\quad \displaystyle\sum_{k=0}^\infty \dfrac{kw^k}{\Gamma(k+\alpha+1)} L_k^\alpha(x)$

$$= e^w (xw)^{-\alpha/2}\{wJ_\alpha(2(xw)^{\frac{1}{2}}) - (xw)^{\frac{1}{2}} J_{\alpha+1}(2(xw)^{\frac{1}{2}})\}.$$

Using this identity and changing variables, we have

(7.5.17) $\quad V_\varepsilon^\alpha f(x) = \dfrac{2e^{x/2}}{\Gamma(\varepsilon + i\theta)} \int_0^\infty \int_0^\infty \int_0^1 (1-u^2)^{\varepsilon-1+i\theta} u^{\alpha+3}$

$$\times \{sJ_{\alpha+2}(suz) - uzJ_{\alpha+3}(suz)\}$$

$$\times g_\varepsilon(z) J_\alpha((2x)^{\frac{1}{2}}s) du dz ds$$

where $g_\varepsilon(z)$ is defined by

$$g_\varepsilon(z) = f\Big(\dfrac{z^2}{2}\Big) e^{-z^2/4} (z^2/2)^{(\varepsilon+i\theta)/2} z.$$

If we let

(7.5.18) $\quad W_\varepsilon^\alpha f(x) = \dfrac{2e^{x/2}}{\Gamma(\varepsilon + i\theta)} \int_0^\infty \int_0^\infty \int_0^1 u^{\alpha+3}(1-u^2)^{\varepsilon-1+i\theta} s$

$$\times J_{\alpha+2}(suz) g_\varepsilon(z) J_\alpha((2x)^{\frac{1}{2}}s) du dz ds,$$

(7.5.19) $$F_\varepsilon^\alpha f(x) = \frac{2e^{x/2}}{\Gamma(\varepsilon+i\theta)} \int_0^\infty \int_0^\infty \int_0^1 u^{\alpha+4}(1-u^2)^{\varepsilon-1+i\theta} z$$
$$\times J_{\alpha+3}(suz) g_\varepsilon(z) J_\alpha((2x)^{\frac{1}{2}}s) du\, dz\, ds,$$

then we have
$$V_\varepsilon^\alpha f(x) = W_\varepsilon^\alpha f(x) - F_\varepsilon^\alpha f(x)$$
since the iterated integrals in $W_\varepsilon^\alpha f$ and $F_\varepsilon^\alpha f$ are finite.

Now we use the identity

(7.5.20) $$J_\alpha((2x)^{\frac{1}{2}}s) = 2(\alpha+1)(2x)^{-\frac{1}{2}} s^{-1} J_{\alpha+1}((2x)^{\frac{1}{2}}s)$$
$$- J_{\alpha+2}((2x)^{\frac{1}{2}}s)$$

to express $W_\varepsilon^\alpha f$ as the sum of two terms. Let

(7.5.21) $$D_\varepsilon^\alpha f(x) = \frac{4(\alpha+1)e^{x/2}}{\Gamma(\varepsilon+i\theta)(2x)^{\frac{1}{2}}} \int_0^\infty \int_0^\infty \int_0^1 u^{\alpha+3}(1-u^2)^{\varepsilon-1+i\theta}$$
$$\times J_{\alpha+2}(suz) g_\varepsilon(z) J_{\alpha+1}((2x)^{\frac{1}{2}}s) du\, dz\, ds,$$

(7.5.22) $$E_\varepsilon^\alpha f(x) = \frac{2e^{x/2}}{\Gamma(\varepsilon+i\theta)} \int_0^\infty \int_0^\infty \int_0^1 u^{\alpha+3}(1-u^2)^{\varepsilon-1+i\theta} s$$
$$\times J_{\alpha+2}(suz) g_\varepsilon(z) J_{\alpha+2}((2x)^{\frac{1}{2}}s) du\, dz\, ds.$$

Then we have $W_\varepsilon^\alpha f(x) = D_\varepsilon^\alpha f(x) - E_\varepsilon^\alpha f(x)$ so that

(7.5.23) $$V_\varepsilon^\alpha f(x) = D_\varepsilon^\alpha f(x) - E_\varepsilon^\alpha f(x) - F_\varepsilon^\alpha f(x).$$

We observe that $E_\varepsilon^\alpha f = G_\varepsilon^{\alpha+2} f$ and hence as $\alpha + 2 \geq 0$ we obtain

(7.5.24) $$\|E_\varepsilon^\alpha f\|_p \leq M(\theta)\left(\|f(x)x^{\varepsilon/2}\|_p + \|f(x)x^{-\varepsilon/2}\|_p\right),$$

(7.5.25) $$\|E_\varepsilon^\alpha f\|_{p,w} \leq M(\theta)\left(\|f(x)x^{\varepsilon/2}\|_{p,w} + \|f(x)x^{-\varepsilon/2}\|_{p,w}\right).$$

In order to estimate $D_\varepsilon^\alpha f$ and $F_\varepsilon^\alpha f$ we obtain the following integral representations.

Proposition 7.5.1.

(i) $D_\varepsilon^\alpha f(x) = \dfrac{(\alpha+1)}{\Gamma(\varepsilon+1+i\theta)} \displaystyle\int_x^\infty f(y) e^{-(y-x)/2} y^{(\varepsilon+i\theta)/2} (\dfrac{x}{y})^{\alpha/2}$

$\times (1 - \dfrac{x}{y})^{\varepsilon+i\theta} y^{-1} dy,$

(ii) $F_\varepsilon^\alpha f(x) = \dfrac{(\alpha+1)}{\Gamma(\varepsilon+1+i\theta)} \displaystyle\int_x^\infty f(y) e^{-(y-x)/2} S_\varepsilon^\alpha(x,y) y^{(\varepsilon+i\theta)/2} (\dfrac{x}{y})^{\alpha/2} dy$

where $S_\varepsilon^\alpha(x,y)$ is the function defined by

$$S_\varepsilon^\alpha(x,y) = \left(1 - \dfrac{x}{y}\right)^{\varepsilon+i\theta} \left\{ 1 - \dfrac{\alpha+2}{\alpha+1}\dfrac{x}{y} - \dfrac{\varepsilon+i\theta}{\varepsilon+1+i\theta}\left(1 - \dfrac{x}{y}\right) \right\}.$$

Proof: We will establish (ii); proving (i) is similar. We write $F_\varepsilon^\alpha f(x)$ as

(7.5.26) $F_\varepsilon^\alpha f(x) = \displaystyle\lim_{\lambda \to 0} \dfrac{2e^{x/2}}{\Gamma(\varepsilon+i\theta)} \int_0^\infty \int_0^\infty \int_0^1 u^{\alpha+4}(1-u^2)^{\varepsilon-1+i\theta} z$

$\times J_{\alpha+3}(suz) g_\varepsilon(z) J_\alpha((2x)^{\frac{1}{2}} s) s^{-\lambda} du\,dz\,ds.$

The factor $s^{-\lambda}$, $0 < \lambda < 1$, enables us to invert the order of integration in the above iterated integral. If $R_\lambda(u,x)$ is the kernel defined in (7.3.7) we can write

(7.5.27) $F_\varepsilon^\alpha f(x) = \displaystyle\lim_{\lambda \to 0} \dfrac{2e^{x/2}}{\Gamma(\varepsilon+i\theta)} \int_0^1 \int_0^\infty u^{\alpha+4}(1-u^2)^{\varepsilon-1+i\theta}$

$\times g_\varepsilon(z) z R_\lambda(uz, (2x)^{\frac{1}{2}}) du\,dz.$

The estimates (i) and (ii) of Lemma 7.3.2 enable us to take the limit under the integral sign. Using (iii) and (iv) of Lemma 7.3.2 one sees that

(7.5.28) $F_\varepsilon^\alpha f(x) = \dfrac{2^{1+\alpha/2}(\alpha+1)e^{x/2} x^{\alpha/2}}{\Gamma(\varepsilon+i\theta)} \displaystyle\int_0^1 \int_{(2x)^{\frac{1}{2}} u^{-1}}^\infty u^3 (1-u^2)^{\varepsilon-1+i\theta}$

$\times g_\varepsilon(z) z^{-\alpha} \left(1 - \dfrac{2(\alpha+2)}{\alpha+1} x u^{-2} z^{-2}\right) dz\,du$

$= \dfrac{(\alpha+1)e^{x/2}}{\Gamma(\varepsilon+i\theta)} \displaystyle\int_0^1 \int_{x/v}^\infty v(1-v)^{\varepsilon-1+i\theta} f(y) e^{-y/2}$

$\times y^{(\varepsilon+i\theta)/2} (\dfrac{x}{y})^{\alpha/2} \left(1 - \dfrac{\alpha+2}{\alpha+1}\dfrac{x}{vy}\right) dy\,dv.$

Changing the order of integration, and performing the v integral we obtain the expression (ii) for $F_\varepsilon^\alpha f(x)$. This completes the proof of the proposition. ∎

We can now get the desired estimates for $D_\varepsilon^\alpha f$ and $F_\varepsilon^\alpha f$. As the function $S_\varepsilon^\alpha(x,y)$ is bounded by a constant C independent of x, y and ε we have

$$(7.5.29) \quad |F_\varepsilon^\alpha f(x)| \leq \frac{C}{|\Gamma(\varepsilon + 1 + i\theta)|} \int_x^\infty |f(y)| y^{\varepsilon/2} \left(\frac{x}{y}\right)^{\alpha/2} e^{-(y-x)/2} dy.$$

Let K be the convolution operator

$$(7.5.30) \quad Kf(x) = \int_x^\infty f(y) e^{-(y-x)/2} dy$$

which is given by the regular kernel $k(u) = e^{u/2} \mathcal{X}_{(-\infty,0)}(u)$. When $\alpha \geq 0$ it follows that

$$(7.5.31) \quad |F_\varepsilon^\alpha f(x)| \leq M(\theta) K \widetilde{f}(x)$$

where $\widetilde{f}(x) = |f(x)| x^{\varepsilon/2} \mathcal{X}_{(0,\infty)}(x)$. As K is a bounded operator on $L^p(\mathbb{R})$, $1 < p < \infty$ it follows that $\|F_\varepsilon^\alpha f\|_p \leq M(\theta) \|f(x) x^{\varepsilon/2}\|_p$. Since the weight $w(x) = x^{p/4-1/2}$ belongs to A_p it also follows that $\|F_\varepsilon^\alpha f\|_{p,w} \leq M(\theta) \|f(x) x^{\varepsilon/2}\|_{p,w}$. Thus we have the desired estimates for F_ε^α when $\alpha \geq 0$.

When $-1 < \alpha < 0$ we dominate $|F_\varepsilon^\alpha f(x)|$ by

$$(7.5.32) \quad |F_\varepsilon^\alpha f(x)| \leq M(\theta) K f_\varepsilon^\alpha(x) |x|^{\alpha/2}$$

where $f_\varepsilon^\alpha(x) = |f(x)| x^{(\varepsilon-\alpha)/2} \mathcal{X}_{(0,\infty)}(x)$. Since $|x|^{\alpha p/2}$ belongs to A_p for $\left(1 + \frac{\alpha}{2}\right)^{-1} < p < -\frac{2}{\alpha}$ we get

$$(7.5.33) \quad \|F_\varepsilon^\alpha f\|_p \leq M(\theta) \left(\int_{-\infty}^\infty |f(x)|^p x^{(\varepsilon-\alpha)p/2} x^{\alpha p/2} dx \right)^{\frac{1}{p}}$$
$$= M(\theta) \|f(x) x^{\varepsilon/2}\|_p.$$

When $\alpha \geq -\frac{1}{2}$ the function $|x|^{\alpha p/2 + p/4 - 1/2}$ is in A_p, $1 < p < \infty$. Hence we also have the weighted inequality

$$(7.5.34) \quad \|F_\varepsilon^\alpha f\|_{p,w} \leq M(\theta) \|f\|_{p,w}, \quad \alpha \geq -\frac{1}{2}, \ 1 < p < \infty.$$

Thus we have the desired estimates for $F_\varepsilon^\alpha f$.

Finally we consider $D_\varepsilon^\alpha f$. From the integral representation (i) of Proposition 7.5.1 we have

$$(7.5.35) \qquad |D_\varepsilon^\alpha f(x)| \leq M(\theta) \int_x^\infty |f(y)| y^{\varepsilon/2} \left(\frac{x}{y}\right)^{\alpha/2} y^{-1} dy.$$

When $\alpha p/2 + 1 > 0$ it follows by weighted Hardy's inequality that

$$(7.5.36) \qquad \|D_\varepsilon^\alpha f(x)\|_p^p \leq M(\theta)^p \int_0^\infty \left(\int_x^\infty |f(y)| y^{(\varepsilon-\alpha)/2} y^{-1} dy\right)^p x^{\frac{\alpha p}{2}} dx$$

$$\leq M(\theta)^p \int_0^\infty |f(y)|^p y^{(\varepsilon-\alpha)p/2} y^{\alpha p/2} dy$$

which gives the inequality

$$(7.5.37) \qquad \|D_\varepsilon^\alpha f\|_p \leq M(\theta) \|f(x) x^{\varepsilon/2}\|_p.$$

When $\alpha \geq -\frac{1}{2}$, $\frac{p}{4} + \alpha\frac{p}{2} + \frac{1}{2} \geq 0$ and hence we also have the inequality

$$(7.5.38) \qquad \|D_\varepsilon^\alpha f\|_{p,w} \leq M(\theta) \|f(x) x^{\varepsilon/2}\|_{p,w}, \quad 1 < p < \infty.$$

Thus we have proved the inequality

$$\|V^\alpha f\|_p \leq M(\theta) \|f\|_p$$

for $\alpha \geq 0$, $1 < p < \infty$ or $-1 < \alpha < 0$, $\left(1 + \frac{\alpha}{2}\right)^{-1} < p < -\frac{2}{\alpha}$. Also we have the weighted inequality

$$\|V^\alpha f\|_{p,w} \leq M(\theta) \|f\|_{p,w}$$

for $\alpha \geq -\frac{1}{2}$, $1 < p < \infty$. Putting the estimates for U^α and V^α together we have got the estimates for $\widetilde{T}_\alpha^{\alpha+2+i\theta} f$. This together with the estimates for $\widetilde{T}_\alpha^{\alpha+i\theta}$ completes the proof of Proposition 7.2.2. Hence both transplantation theorems are proved. ∎

BIBLIOGRAPHY

[A] R. Askey, *Orthogonal polynomials and positivity*, In: Studies in Applied Mathematics, Wave propagation and special functions, SIAM (1970), 64–85.

[AW] R. Askey and S. Wainger, *Mean convergence of expansions in Laguerre and Hermite series*, Amer. J. Math. **87** (1965), 695–708.

[BNT] P. L. Butzer, R. J. Nessel and W. Trebels, *On summation processes of Fourier expansions in Banach spaces*, Tôhoku Math. J. **24** (1972), 127–140.

[CC] S. Chanillo and M. Christ, *Weak $(1,1)$ bounds for oscillating singular integrals*, Duke Math. J. **55**, No. 1 (1987), 141–155.

[Dl] J. Dlugosz, L^p *multipliers for Laguerre expansions*, Colloq. Math. **54** (1987), 287–293.

[EM1] A. Erdelyi, W. Magnus, F. Oberhettinger and F. G. Tricomi, *Higher transcendental functions*, McGraw Hill, New York, 1953.

[EM2] A. Erdelyi, W. Magnus, F. Oberhettinger and F. G. Tricomi, *Tables of integral transforms*, vol. I, McGraw Hill, New York, 1954.

[Fe1] C. Fefferman, *The multiplier problem for the ball*, Ann. Math. **99** (1971), 330–336.

[Fe2] C. Fefferman, *A note on spherical summation multipliers*, Israel J. Math. **15** (1973), 44–52.

[Fo] G. B. Folland, *Harmonic analysis in phase space*, Ann. Math. Stud., Princeton Univ. Press **112** (1989).

[G] L. Garding, *On the asymptotic distribution of eigenvalues and eigenfunctions of elliptic differential operators*, Math. Scand. **1** (1953), 237–255.

[Ge] D. Geller, *Fourier analysis on the Heisenberg group*, J. Funct. Anal. **36** (1980), 205–254.

[GM1] E. Gorlich and C. Markett, *A convolution structure for Laguerre series*, Indag. Math. **44** (1982), 161–171.

[GM2] E. Gorlich and C. Markett, *Mean Cesàro summability and operator norms for Laguerre expansions*, Comment. Math. Prace Mat., Tomus specialis II (1979), 139–148.

[Gn] J. J. Gergen, *Summability of double Fourier series*, Duke Math. J. **3** (1937), 133–148.

[Go] D. Goldberg, *A local version of real Hardy spaces*, Duke Math. J. **46** (1979), 27–42.

[GS] G. Szego, *Orthogonal polynomials*, Amer. Math. Soc. Colloq. Pub. 23, Providence, R.I. (1967).

[HJ] A. Hulanicki and J. Jenkins, *Almost everywhere summability on nilmanifolds*, Trans. Amer. Math. Soc. **278** (1983), 703–715.

[K] Y. Kanjin, *A transplantation theorem for Laguerre series*, Tôhoku Math. J. **43** (1991), 537–555.

[KST] C. Kenig, R. Stanton and P. Tomas, *Divergence of eigenfunction expansions*, J. Funct. Anal. **46** (1982), 28–44.

[KS] K. Stempak, *Almost everywhere summability of Laguerre series*, Stud. Math. **100 (2)** (1991), 129–147.

[M] G. Mauceri, *The Weyl transform and bounded operators on $L^p(\mathbb{R}^n)$*, J. Funct. Anal. **39** (1980), 408–429.

[Ma1] C. Markett, *Norm estimates for Cesàro means of Laguerre expansions, Approximations and function spaces*, (Proc. Conf. Gdansk, 1979), North Holland, 1981, pp. 419–435.

[Ma2] C. Markett, *Mean Cesàro summability of Laguerre expansions and norm estimates with shifted parameter*, Analysis Math. **8** (1982), 19–37.

[Mc] J. McCully, *The Laguerre transform*, SIAM Rev. **2** (1960), 185–191.

[Mu1] B. Muckenhoupt, *Mean convergence of Hermite and Laguerre series I*, Trans. Amer. Math. Soc. **147** (1970), 419–431.

[Mu2] B. Muckenhoupt, *Mean convergence of Hermite and Laguerre series II*, Trans. Amer. Math. Soc. **147** (1970), 433–460.

[P] J. Peetre, *Remark on eigenfunction expansions for elliptic operators with constant coefficients*, Math. Scand. **15** (1964), 83–92.

[Po] E. Poiani, *Mean Cesàro summability of Hermite and Laguerre series*, Trans. Amer. Math. Soc. **173** (1972), 1–31.

[RS] F. Ricci and E. Stein, *Harmonic analysis on nilpotent Lie gorups and singular integrals*, J. Funct. Anal. **73** (1987), 179–194.

[S1] E. Stein, *Singular integrals and differentiability properties of functions*, Princeton University Press (1971).

[S2] E. Stein, *Topics in harmonic analysis related to Littlewood-Paley theory*, Ann. Math. Stud., Princeton University Press **63** (1971).

[S3] E. Stein, *Beijing lectures on Fourier analysis*, Ann. Math. Stud., Princeton University Press **112** (1986).

[So] C. Sogge, *On the convergence of Riesz means on compact manifolds*, Ann. Math. **126** (1987), 439–447.

[SW] E. Stein and G. Weiss, *Introduction to Fourier analysis on Euclidean spaces*, Princeton University Press, 1981.

[St1] R. Strichartz, *Convolution with kernels having singularities on a sphere*, Trans. Amer. Math. Soc. **148** (1970), 461–471.

[St2] R. Strichartz, *Harmonic analysis as spectral theory of Laplacians*, J. Funct. Anal. **87** (1989), 51–148.

[T1] S. Thangavelu, *Multipliers for Hermite expansions*, Revist. Mat. Ibero. **3** (1987), 1–24.

[T2] S. Thangavelu, *Summability of Hermite expansions I*, Trans. Amer. Math. Soc. **314** (1989), 119–142.

[T3] S. Thangavelu, *Summability of Hermite expansions II*, Trans. Amer. Math. Soc. **314** (1989), 143–170.

[T4] S. Thangavelu, *Summability of Laguerre expansions*, Analysis Math. **16** (1990), 303–315.

[T5] S. Thangavelu, *On almost everywhere and mean convergence of Hermite and Laguerre expansions*, Colloq. Math. **60** (1990), 21–34.

[T6] S. Thangavelu, *Riesz transforms and the wave equation for the Hermite operator*, Comm. p.d.e. **15** (8) (1990), 1199–1215.

[T7] S. Thangavelu, *Littlewood-Paley-Stein theory on $\mathbb{C}^n$ and Weyl multipliers*, Revist. Mat. Ibero. **6** (1990), 75–90.

[T8] S. Thangavelu, *Weyl multipliers, Bochner-Riesz means and special Hermite expansions*, Ark. Mat. **29** (1991), 307–321.

[T9] S. Thangavelu, *Hermite expansions on $\mathbb{R}^n$ for radial functions*, Proc. Amer. Math. Soc. (to appear).

[T10] S. Thangavelu, *Transplantation, summability and multipliers for multiple Laguerre expansions*, Tôhoku Math. J. (to appear).
[T11] S. Thangavelu, *On conjugate Poisson integrals and Riesz transforms for the Hermite expansions* (to appear in Colloq. Math.).
[Ta] M. Taylor, *Non commutative harmonic analysis*, Amer. Math. Soc., Providence, R.I. (1986).
[W1] G. N. Watson, *Another note on Laguerre polynomials*, J. London Math. Soc. **14** (1939), 19–22.
[W2] G. N. Watson, *A treatise on the theory of Bessel functions*, Cambridge University Press, London, 1966.